Economic and Financial Analyses of Small and Medium Food Crops Agro-Processing Firms in Ghana

The United Nations University Institute for Natural Resources in Africa (UNU-INRA), is one of the 15 Research and Training Centres/ Programmes (RTC/Ps) that constitute the United Nation's University's (UNU) worldwide network. The Institute's mission is to empower African universities and other research institutions through capacity strengthening. UNU-INRA operates mainly from its headquarters in Accra and Operating Units (OUs) currently based at universities in Cameroon, Cote d'Ivoire, Namibia, Senegal and Zambia.

United Nations University Institute for Natural Resources in Africa (UNU-INRA)
Second floor, International House, Annie Jiagge Road,
University of Ghana, Legon Campus
Accra, Ghana

Private Mail Bag, Kotoka International Airport,
Accra, Ghana

Email: *inra@unu.edu or unuinra@gmail.com*
www.inra.unu.edu

ECONOMIC AND FINANCIAL ANALYSES OF SMALL AND MEDIUM FOOD CROPS AGRO-PROCESSING FIRMS IN GHANA

Authors

Timothy Afful-Koomson
William Fonta
Stephen Frimpong
Nathaniel Amoh

Economic and Financial Analyses of Small and Medium Food Crops Agro-Processing Firms in Ghana

Authors
Timothy Afful-Koomson
William Fonta
Stephen Frimpong
Nathaniel Amoh

ISBN: 978-9988-633-85-1

Editing and Layout: Praise Nutakor

Cover Design: Kwabena O. Asubonteng

Printed by: Pixedit Limited, Ghana
+233-203339269/+233-206893271

TABLE OF CONTENTS

LIST OF TABLES

LIST OF FIGURES

FOREWORD

Agriculture is the main driving force of Ghana's economy. The agriculture sector accounts for one-quarter of GDP and employs about 42 percent of the country's work force. Food processing is predominantly dominated by small to medium scale enterprises usually in the informal sector.

One of the biggest challenges of the sector is post-harvest losses due to non-availability of efficient food processing technologies, storage facilities and the inability of our farmers to readily have access to markets. As a government, we have put in place various measures including the construction of storage facilities in some communities and the provision of loan facilities to farmers through government initiatives such as the Export Development and Agricultural Investment Fund (EDAIF), the Venture Capital Trust Funds (VCTF), Microfinance and Small Loans Centre (MASLOC) and the Ghana Agricultural Insurance Program (GAIP), in order to enhance efficiency in the sector. Despite several government interventions, the country still continues to import some food products such as rice, meat and poultry products because of low productivity.

I find this book *"Economic and Financial Analyses of Small and Medium Food Crops Agro-Processing Firms in Ghana"* very relevant for the sector's development. This is because, this book, which reports on a research project that analysed agro-processing firms' economic and financial situation in Ghana, provides the empirical evidence for us stakeholders, to re-think our strategic direction to improve food crops agro-processing in Ghana.

This monograph is particularly important for the sector's development because it focuses on the country's major food crops, which are cereals and starchy foods. It also provides us with scientific knowledge that reiterates major constraints facing agro-processing firms in the country. These challenges include standardization, quality control, packaging, marketing and access to credit facilities. This reinforces the need for public-private partnership to enhance the capacity of food crops agro-processing firms in packaging, promoting and marketing their products to help reduce the influx of imports of processed food products into the country.

The publication also emphasizes agro-processing firms' inability to purchase appropriate technologies to improve productivity, due to financial constraints. This empirical endorsement of the numerous challenges confronting the sector will enable government and other stakeholders in the agricultural value chain to re-direct efforts meant to improve the sector's performance.

I believe the information provided in this book will add to existing knowledge available to government in formulating policies to improve productivity of the sector. It will also provide investment opportunities for non-state actors in the agro-processing sector to take advantage of the numerous opportunities in the sector to drive industrialization, improve export and increase employment opportunities for the country's development.

I thank the United Nations University - Institute for Natural Resources in Africa (UNU-INRA) for supporting this research project to help both state and non-state actors in the agricultural sector to make informed decisions in increasing productivity of food crops agro-processing firms in Ghana.

Honourable Clement Kofi Humado

Minister of Food and Agriculture, Ghana

(June, 2014)

ACKNOWLEDGEMENTS

This study received financial support from the Ministry of Agriculture, Forestry and Fisheries of the Government of Japan and the United Nations University – Institute for Peace and Sustainability (UNU-ISP) through the On-the-Job Research Capacity Building (OJCB). We are grateful for the financial support through the OJCB that enabled Dr William Fonta to participate in this study.

Special thanks also go to Honorable Clement Kofi Humado, the Minister of MOFA, Honorable Yaw Effah-Baafi, the Deputy Minister for Crops, and Mr Maurice Tanco Abisa-Seidu, the Chief Director of MOFA, for their suggestions and support in gaining access to critical information and data for the study.

We also thank Mr. Emmanuel Asante Krobea, the Director of the Crop Services Directorate, Regional and District Directors of MOFA in the Brong Ahafo, Central, Eastern, Northern and Western Regions for the assistance provided during the field study and data collection.

We appreciate the comments and other inputs received from participants at the December 4th, 2013 Stakeholders Seminar where the preliminary findings were discussed. Thanks to representatives from key stakeholders including Ms. Sitsofe Ama Gadzanku (Ministry of Trade and Industry); Ms. Charlotte Oppong Baah (Ministry of Food and Agriculture); Mr Michael W. Awuku (Export Development and Agric Investment Fund); Mr James Nsiah (Venture Capital Trust Fund); Mr Shadrack Quarcoo (MASLOC); Mr. Kenneth Key (International Finance Corporation); Mr. Enno Heijndermans (SNV-Ghana); Mr Ransford Kofi Kani (NBSSI); Mr Kwame A. Attrams (Agric Development Bank); Ms. Anna Mensah (GEF/SCP-UNDP); Mr. Dakurah Eugene Yinou (Ministry of Local Government and Rural Development); Mr. Kingsley Adofo-Addo (ECOBANK-Ghana); Mr Kwakye Ameyaw and Ms Diana Fiati (Forestry Commission); Prof. Daniel B. Sarpong and Dr. John K.M. Kuwornu (College of Agric. and Consumer Sciences, University of Ghana).

We also want to thank Nana Kwame Appiah and Belinda Otu, Interns of UNU-INRA who supported us during the field work. We are also grateful to Mr. William Briandt, the Driver at UNU-INRA who coordinated the schedules with the District Directors of MOFA and the transportation for the field work.

ACKNOWLEDGEMENTS

This study received financial support from the [illegible] of [illegible] [illegible] of the [illegible] and [illegible] Institute for Peace and Security [illegible] (UNU-INRA?) through the [illegible] [illegible] (OSIWA?). We are grateful for the [illegible] through the [illegible] that enabled the [illegible] [illegible] [illegible].

[illegible] to Honourable [illegible], the [illegible] of [illegible] and [illegible] [illegible] Department of [illegible] [illegible] Mr [illegible] Abbey [illegible] Chief [illegible] [illegible] the [illegible] and [illegible] for [illegible] and [illegible].

We [illegible] thank [illegible] [illegible] Regional and District [illegible] [illegible] [illegible] Northern and [illegible] [illegible] [illegible] [illegible] [illegible].

We appreciate the [illegible] and [illegible] inputs received [illegible] [illegible] [illegible] December [illegible] [illegible] [illegible] [illegible] [illegible] [illegible] including [illegible] [illegible] [illegible] [illegible] [illegible] [illegible] [illegible] Ministry of [illegible] and [illegible] [illegible] [illegible] [illegible] [illegible] [illegible] [illegible] [illegible] [illegible] [illegible] [illegible] Development [illegible] [illegible] [illegible] [illegible] Ministry of Local Government and [illegible] [illegible] [illegible] [illegible] [illegible] [illegible] [illegible] [illegible] [illegible] [illegible] Sciences, University of Ghana.

We also want to thank [illegible] [illegible] [illegible] [illegible] [illegible] who supported in [illegible] [illegible] [illegible] [illegible] Mr William Baaode, the University [illegible] [illegible] who [illegible] [illegible] [illegible] [illegible] [illegible] [illegible] [illegible] [illegible] [illegible] [illegible].

EXECUTIVE SUMMARY

Agriculture still remains the bedrock for most developing countries' economies. In Africa, it supports the livelihoods of more than 80 percent of the population and provides employment for close to 60 percent of the economically active population. It has been shown that growth which is stimulated by agriculture is twice as effective in reducing poverty as growth based on other sectors in Africa. One way of stimulating agricultural-based growth in Africa is to avoid losses along the agricultural value chain. It is estimated that about 50 percent of perishable food commodities such as fruits, vegetables, roots and tubers are lost every year through, for example spoilage, damage and loss of nutritional values. Post-harvest losses (PHL's) imply more food insecurity, and widening poverty.

In Ghana, post-harvest losses in Maize, Cassava, Yam and Rice amounts to over 35.1 percent, 34.6 percent, 24.4 percent and 6.9 percent respectively [Ministry of Food and Agriculture (MoFA), 2007]. These losses are as a result of ineffective food processing technologies, careless harvesting and inefficient post-harvest handling practices. It is also due to damages during transportation because of bad roads, inappropriate market practices and inadequate storage facilities. An evaluation of the Root and Tuber Improvement Programs (RTIMP) by MoFA, revealed that most sweet potato farmers have not increased their incomes as a result of the limited options for processing and marketing their produce.

Thus, apart from reducing PHL's agro-processing industry is also very important to improving sustainable local food supply, employment, income and inclusive growth in the nation. According to FAO (2012), food security is significantly related to the efficient processing and distribution of agricultural products including food crops. Sadly, only five (5) per cent of food products traded in Ghana is processed, and the total volume of processed food crops exported is relatively low (MOFA 2005; 2012). The country is pursuing some import substitution strategies to control the importation of certain food products (GNA, 2014). The outcome from these initiatives such as the import control of rice will be very modest without complementary initiatives to improve processing, storage, packaging and to promote quality locally-manufactured rice that will be competitive to imported rice products. Improved agro-processing of food crops also presents huge opportunities for the country to promote economic value addition, resource use efficiency, rural industrialization, agribusiness enterprise development, employment, alternative income and inclusive growth in Ghana.

In this study, economic and financial analyses of micro, small and medium food crops agro-processing firms in Ghana are carried out. The central goal of the study is to provide a solid knowledge base for improving efficiencies in food crops agro-processing activities in general and to raise sector-wide food crop agro-processing operations in Ghana.

The book is organized into five major parts. Part I includes Chapters 1 and 2. Part II is made up of Chapters 3 to 5 and Part III includes Chapters 6 to 8. Part IV is made up of Chapters 9 to 12 and Part V is made up of Chapter 13.The first chapter presents the context, opportunities and challenges for promoting food crops agro-processing in Ghana. The intent is to provide a cogent and coherent insight into the persistent low level of food crop agro-processing in Ghana. It is also intended to demonstrate the urgency of evidence-base knowledge to inform policy and decision-making to take advantage of the enormous opportunities for using agro-processing to rationalize industrialization, import substitution, job creation and inclusive growth in the nation. Chapter two describes the research methodology used in the study as well as the theoretical basis of the study. In order to have the diversity of food crops processed as well as the diversity of size, scale and scope of the agro-processing firms in the study, Ghana was stratified into major producing regions for cereal (maize, millet, sorghum, and rice), tubers and starchy staples (cassava, yam and plantain), legumes (groundnuts, cowpea and soybean) and fruits and vegetables. The regions were then categorized into major food crops produced based on data from MOFA (2012) on estimated metric tons of production. Due to the predominance of regions in the tubers categories, Brong Ahafo and Western Regions were randomly selected. Eastern Region was selected for fruits and vegetables, Central Region for cereals and Northern Region for legumes. Overall, 272 firms were randomly sampled for the study.

In chapter three, a description of the agro-processing sector in Ghana is presented as well as some of the major constraints facing the sector. The agro-processing sector in Ghana is largely dominated by horticultural products (fruits and beverage), processing of vegetables such as okra, tomatoes, pepper, onions, cabbage, lettuce, green beans and mushrooms, processing of cereals (maize and rice), processing of roots and tubers and processing of oil palm. Amongst the constraints facing the sector, lack of access to appropriate technology is foremost. Most agro-industries in Ghana operate using indigenous technologies. This is as a result of the high cost of processing equipment and the limited capacity of firms to mobilize capital to purchase labor saving technologies especially when it has to be imported. This makes agro-processing operations time consuming, labor intensive with limited opportunity for scaling up. The reliance on rudimentary technology reduces efficiency and the productivity of operations. It also poses some risks to the health of the workers in the firm. Inadequate financing and credit is also a long standing problem affecting the development of the small scale industries.

Chapter four focuses on the measurement of concentration and distribution of food crops agro-processing firms in Ghana. This is intended to provide information on the extent of competition and concentration in the food crop agro-processing sector. This information is very useful for business policy and further development of the sector. Firstly, the results indicate that more than 85 percent of agro-processing firms in Ghana are micro-enterprises. This finding supports NBSSI (1991) findings that about 90% of firms in Ghana are micro or small scale enterprises. Very small firms which employ about six to nine people constitute 7%. Small agro-processing firms form only 5% whereas medium agro-processing firms form 3%. Secondly, the results further indicates that agro-processing in Ghana are skewed towards two major product categories- oil, and root and tuber processing. Palm oil producing agro-processing firms account for more than half of all agro-processing firms in Ghana. The skewness of agro-processing towards particular agro-products may be influenced by demand or availability of market for the products. Also, the Root and Tuber Improvement Programme and the relatively ready internal or international market for roots and tuber products such as 'gari' may be some of the reasons for the high percentage of firms focusing on these products.

Finally, the Western, Central and the Northern regions of Ghana depict low concentration of food crop agro-processing firms while the Eastern and Brong-Ahafo regions reflect relatively higher concentration of food crop agro-processing firms using concentration curves. We found high consistency with calculations based on the Herfindahl-Hirschman index (HHI).

The major ownership and employment characteristics of food crops agro-processing firms in Ghana are treated in chapter five while chapter six explores the operation of food crops agro-processing firms in General. This includes their installed and operating capacities, the level of technology use, efficiency of conversion, internal and external linkages of operation, linkages with input suppliers and financial institutions, relationship with service providers, operations management-repairs and maintenance of fixed assets, and the operations constraints of food crops agro-processing firms. The summary message of chapter six is that the potential transformative role that agro-processing firms could play in Ghana has not been realized due to several critical operation constraints facing them. Several of the firms interviewed are operating below recommended capacity with indigenous and inefficient technologies. The study also established that many of these food crop agro-processing firms have no or weak internal and external linkages. They lack adequate human resource skills as well as operating capital to manage the growth and profitability of the firms. Although most of the firms have the intention to expand, their greater reliance on credit as the major source of funding (which is usually unavailable to them) has limited their business expansion.

Chapter 7 explores the productivity and efficiency of agro-processing firms in Ghana. Different regressions are estimated for the firms in various regions, their scale of production and the type of crop they produce. The Cobb-Douglas production function is estimated using the ordinary least square technique and the various returns to scale are determined. The results suggest that **cost of labor, cost of physical capital and cost of raw materials are significant factors to output sales for food crops agro-processing firms in Ghana**. Cost of raw materials is an important factor explaining sales of output irrespective of the location of the firm, type of crop processed and the scale of production. The cost of labor is also an important factor for firms in the Northern and Brong-Ahafo regions.

It is also a significant factor for firms producing at micro and small scale level and firms who process oil and oil by products. The results also indicate that due to some factors such as low skill of labor, underemployment and the predominance of labor-intensive and rudimentary processing technology used particularly in the Northern Region there is negative marginal productivity of labor for food crops agro-processing firms in this region. Furthermore, the results indicate that **productivity and efficiency of utilizing raw materials improves consistency with scale of operation**. As firms expand from micro, very small to small scale, they become more diversified in their products delivery which also improves their efficiency of conversion of raw materials. The sum of the partial elasticity gives a value of 0.34 which indicates decreasing returns to scale. The returns to scale are increasing for firms in their individual processing category but when assessed in terms of region and scale level the returns to scale tend to decrease.

Chapter 8 highlights the gains from trade in processed food crops as well as the constraints facing firms in the food crop agro-processing industry regarding standardization, quality control and marketing. It was found that although the GSA has standards for food processing, most of the firms in Ghana are not in compliance. We equally found that **for food crops agro-processing firms to improve their competitiveness, they will have to enhance their capacity for packaging, promoting and marketing their products.** They will also have to set in place formal credit agreements with their clients and have effective systems for managing account receivables.

In chapter 9, we examined the capital investments of food crop agro-processing firms in Ghana. The results indicate that most of the firms surveyed do not carry out inventory due to lack of capital, absence of raw materials and availability of space and facilities for storing the inventory. For those carrying out inventory, many do not have efficient inventory management systems because of poor coordination between production and marketing departments. Furthermore, **there is increasing capital efficiency as firms expand their scale of operation.** Micro-, very small-scale firms have poor capital efficiency. This may be due to the deteriorating and rudimentary nature of assets used by most of the firms operating at these scales.

Chapter 10 examines some major financial ratios for assessing the liquidity and profitability of food crops agro-processing firms. We found an **unreasonably high current ratio for most of the firms operating**

at lower scales. For example, micro-scale firms have current ratio above 7. Consistent with normal business practice, current ratio above 3 may signify that the firm is not efficient in managing its current assets or is not taking advantage of appropriate short-term leverage strategies. Medium scale firms have current ratios within the generally acceptable range. The high current ratio of micro-scale firms indicates the need to improve the capacity of these firms for efficient use of available credit. The study also found that **firms processing vegetables have negative profit margins**. This may be due to the low capital efficiency and the high costs of processing vegetable products. Similarly, it was noted that although firms processing alcoholic beverages have the relatively higher annual sales, greater percentage of the sales revenue goes to costs of goods sold and other expenses. However, **generally, most of the firms have low profitability ratios**. Some of the firms are just on the threshold of collapse. Because most of them do not have business plans and do not monitor the performance and risk of the business, they just operate on a subsistence basis.

In chapter 11, some of the credit schemes that may be relevant for promoting food crops agro-processing in Ghana are discussed. These include financial institutions such as banks, micro-credit institution and private equity investors as well as relevant credit schemes established by the government such as the Export Development and Agricultural Investment Fund (EDAIF), the Venture Capital Trust Funds (VCTF), Microfinance and Small Loans Centre (MASLOC) and other financial products and services including the Ghana Agricultural Insurance Program (GAIP). The review found that with the exception of EDAIF that specifically targets beneficiaries that include firms in food crops agro-processing, the other schemes are more generic in nature. Similarly, is the high interest rate and the wide interest rate spread points to the relative dysfunctional and prohibitive system of financial intermediation in Ghana. This undermines the competitiveness of food crops agro-processing firms in Ghana.

Chapter 12 focuses on the financial needs of food crops agro-processing firms in Ghana. The results indicate that majority of the firms do not have access to credit, which explains one of the reasons for the poor capital investment and growth of food crops agro-processing firms in Ghana. **The results show that education, size of firm as well as the business assets of the firm significantly determines access to credit**

by food crop agro-processing firms in Ghana. The relatively high transaction cost of servicing so many micro-level firms is constraining their access to credit. They are also characterized by high degree of informality that worsens their perceived high risks, low returns and poor performance of investments. This result substantiates the recommendation in chapter 4 to consolidate particularly the micro- and very small-scale firms to make them operate at viable scale and with adequate business assets to serve as collateral for loans.

Chapter 13 discusses a simple framework to prioritize issues. It also provides some actionable steps and national strategic implementation plan for pursuing the key interventions suggested. It is only through implementation that Ghana can realize the potential of transforming the food crops agro-processing firms into competitive and viable firms to pursue national efforts at import substitution such as the measures for controlling the importation of rice.

It is hoped that such practical measures should move the nation forward in its efforts at promoting agro-processing and agribusiness required to sustain initiatives to control importation of cheap products and to prevent the decline of industrialization in the nation.

CHAPTER ONE: INTRODUCTION

1.1 Background and Context

Agriculture remains the bedrock for economic development in Ghana despite the discovery of oil and gas in commercial quantities in the nation. According to the Ministry of Food and Agriculture (MOFA), the agriculture sector contributes about 35 percent of the Gross Domestic Product (GDP) in Ghana and provides employment for over 65 percent of the population especially in the rural areas (MOFA 2011). Agriculture is also the basis of any policy that aims to improve food security, wealth enhancement and diversified livelihoods in the nation. This book focuses on the food crops subsector which is the major subsector in terms of its contribution to agricultural GDP, employment, rural income and food security.

Figure 1.1 Agricultural Subsectors by GDP (%)

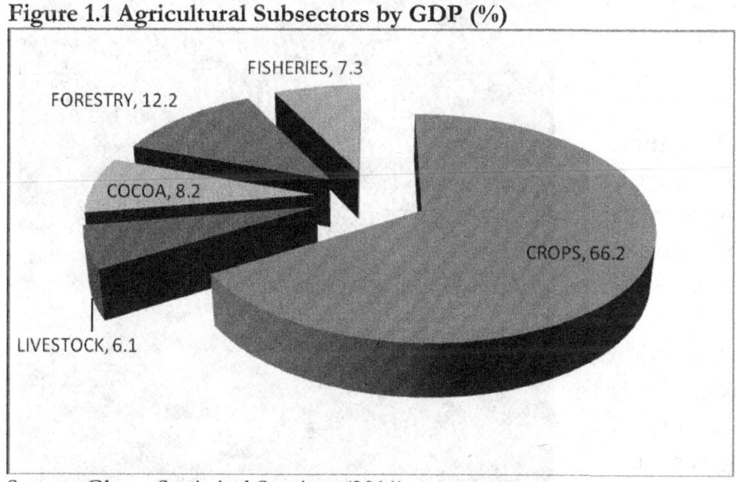

Source: Ghana Statistical Services (2011)

As indicated in Figure 1.1, the food crops subsector accounts for about 66 percent of the total contribution of the agricultural sector to GDP in Ghana. However, this sector is also characterized by high rates of post-harvest losses. It is estimated that about 50 percent of perishable food commodities such as fruits, vegetables, roots and tubers are lost every year through for example spoilage, damage and loss of nutritional values. A relatively lower rate of post-harvest loss of about 30 percent is estimated for food grains such as maize, rice, sorghum, millet and cowpeas (Aworh, 2008; Okorley & Ayekpa, 2010). In 2007, the MOFA estimated the post-harvest losses in Maize, Cassava, Yam and Rice to be 35.1 percent, 34.6 percent, 24.4 percent and 6.9 percent respectively. As indicated in quote 1.1, the post-harvest losses for highly perishable vegetables such as tomatoes could be above 50 percent- especially in areas with low access to transportation.

Quote 1.1. AGOTIME :Ziope 'cries' for tomato factory

Torgbui Binah Lawluvi VI, Paramount Chief of Ziope Traditional Area, has appealed to government to honour its promise of establishing tomato factory at Ziope.

He said the absence of the facility was affecting agriculture in the area since tomato farmers experienced glut every year.

Source: GNA 2013

The chief in Agotime, Volta Region – one of the major producers of tomatoes in the country is requesting the Government for a tomato factory to address the high perishable rate of the produce during the harvesting season (GNA, 2013). These losses are as a result of ineffective food processing technologies, careless harvesting and inefficient post-harvest handling practices. It is also due to damages during transportation because of bad roads, inappropriate market practices and inadequate storage facilities.

Post-harvest losses result in reduced income to farmers. An evaluation of the Root and Tuber Improvement Programs (RTIMP) by MoFA (2010), revealed that most sweet potato farmers have not increased their incomes as a result of the limited options for processing and marketing their produce. So apart from reducing PHL's in the agro-processing industry, it is also very important to improve sustainable local food supply, employment, income and inclusive growth in the nation.

PHL therefore reduces both private and social benefits of food production activities. Food production sectors in Ghana are dominated by small-scale farmers who spend their meager capital and human resources for subsistence farming with low capital and capacity to effectively manage PHL. PHL robs these farmers of the benefits and profits from their hard labor, time, money, energy and resources (World Bank, 2010). PHL constrains farmers' efforts to regulate or smooth food prices and markets due to lack of storage and processing facilities and the urgency to sell their produce before spoilage. Since most farmers have to sell their produce at the same period, there is a period of over-supply and low pricing during harvesting season and period of scarcity afterwards.

PHL is also one of the major factors contributing to worsening food insecurity in Ghana because food lost to pests and insects damage, contamination, infestation by pathogens, spoilage and other physiological deterioration reduce the quantity and quality of food available for consumption. PHL also exacerbates the environmental impacts of agriculture since more land, forests, water and soil have to be used to produce food to compensate for the food wasted through PHL. PHL may also have health implications like the consumption of grain contaminated by aflatoxin (World Bank, 2010). Reducing PHL is therefore very essential not only for improved household income, food security and nutrition but also for food safety and human health. Moreover, exports of agricultural produce from Ghana are dominated by primary commodities which suffer from large price variability and declining world market prices.

1.2 The Role of Agro-processing in Ghana.

Lack of agro-processing and storage facilities are among the major reasons for high PHL, low exports of competitive value-added products and variability of food supply in Ghana. According to MOFA (2005; 2012) only five (5) per cent of food products traded in Ghana is processed, and the total volume of processed food crops exported is relatively low. Enhancing nutritional value and adding economic value through processing are some of the major measures that will contribute to improving food security in Ghana.

The development and introduction of new processing technologies offer potential to improve food security and local industrialization. Processing of agricultural products, therefore, occupies significant position in the agribusiness value chain. Agro-processing improves the efficiency of harvesting, field handling, packaging, storage and marketing of agricultural produce and contributes to prolonging the shelf life of produce thereby reducing spoilage and wastage of food (Plucknett, 1979) (FAO, 1985) (Rolle, 2006).

Quote 1.2.

According to MoFA (2005; 2012), only five (5) per cent of food products traded in Ghana is processed, and the total volume of processed food crops exported is relatively low

Food crops processing has the potential of increasing the market opportunities for agricultural exports since processed goods generally have greater price stability than raw materials (Dijksra, 2001). As recognized in the Ghana Poverty Reduction Strategy II (GPRS II), no significant progress can be made in raising the average real incomes of Ghanaians as a whole without significant improvements in the productivity of the agriculture sector and the agro-based processing industry (NDPC, 2005). It is in light of these that the government set a target to reduce post-harvest losses to an average of 15% by 2015. This is to be achieved by supporting the establishment of food crop agro-processing firms in Ghana (METASSIP, 2012). Thus, in a bid to minimize post-harvest losses and reap the maximum benefit from agriculture, the role of agro-processing has become critical to the economic development of Ghana and security of income and livelihoods of rural communities. Accordingly, many food crop agro-processing firms are being developed in many regions of Ghana (Okorley and Ayekpa, 2009). As a result, increased food crop production is envisaged with proportionate increase or improvement in agro-processing firms in the industry.

1.3 Why Should Ghana focus on Food Crops Agro-processing?

It is appropriate to note that the food crops agro-processing industry in Ghana is dominated by micro, small and medium processing enterprises (MSMEs) using low and rudimentary processing technologies and lowly-skilled labor. They also face other challenges like low quality and irregular supply of production inputs leading to the production of low standard products which are not competitive to imported products. The growing urban food market and increasing demand for high quality and safety standards food especially by the urban middle income class have spurred massive food imports into the country in the absence of competitive local processing companies. The intense competition from cheap food imports is driving many SMEs in the food crops agro-processing industry out of business with consequent effects on reduced opportunities for diversified rural economy, employment, income, and quality standards of living. These unfortunate results also have broader implications for hampering inclusive growth in Ghana.

Among the major reasons why Ghana should focus on improving food crops agro-processing are:

- ***Economic Value Addition – reduce post-harvest losses. Improve resource use efficiency***
- ***Rural industrialization- considering the percentage of the population engaged in farming and primary processing of food crops***
- ***Import substitution – huge competitive advantage in terms of availability of human and natural resources as well as the growing urban population and middle-income class***
- ***Enterprise development***
- ***Employment and improved livelihoods***
- ***Diversification of rural economies and***
- ***Inclusive growth***

Ghana has the capacity to improve agro-processing of food crops in the country. Improved agro-processing will be critically needed to sustain national import substitution measures such as the initiative to control the importation of rice (GNA, 2014). The country could achieve the quality standards and packaging of food crops shown in Figure 1.2. Countries such as Thailand and Vietnam committed significant public and private resources to improving food crops agro-processing and Ghana could do the same.

Figure 1.2 Potential Quality Standards and Packaging of Food Products in Ghana.

The food crops processing industry in Ghana have a great potential for promoting socially-inclusive and competitive development strategies, enhancing economic value addition, improving food supply and in reducing rural poverty in the nation. Enhancing agro-processing of food crops by MSMEs should therefore be one of the priority areas of any national initiative to improve agricultural development and food supply in Ghana. And it will be ridiculous to talk of promoting food crops agro-industries in Ghana without specific initiatives to improve access to financial products and services for MSMEs in agro-processing. Developing innovative and inclusive financial services for MSMEs in agro-processing will enable them to improve their inventory management, quality control and marketing of their processed products.

It is encouraging to note that the country has realized the socio-economic benefits from the export of value added agricultural products and has put in place some initiatives that have contributed to some improvement in the exports of processed and prepared foods. For example, in 2007, total exports of processed foods (excluding processed cocoa) amounted to US $177,639,586 which represents a 17.2 percent increase over the previous year. There are some initiatives such as the Venture Capital Trust Funds (VCTF), Microfinance and Small Loans Centre (MASLOC) and the Export Development and Agricultural Investment Fund (EDAIF). However, there is the critical need for a coordinated and comprehensive national strategy to promote access to financial products and services for MSMEs in food crops agro-processing.

Ghana needs an integrated and strategic national implementation plan for food crops agro-industries that is built on these initiatives outlined above, national policies such as the Food and Agriculture Sector Development Policy (FASDEP II) and regional initiatives such as the Comprehensive Africa Agriculture Development Program (CAADP) and the African Agribusiness and Agro-industries Development Initiative (3ADI). Without a comprehensive and strategic implementation plan that takes into consideration the specific characteristics of MSMEs in this subsector of the agribusiness value chain, the outcome from the disjointed initiatives could be minimal. Such national strategic implementation plan could also focus on coordinating the outcomes from the different initiatives including regional initiatives such as the 3ADI that advocates for increased private sector investment flows for agribusiness and agro-industrial development (FAO and UNIDO, 2010).

1.4. The objectives and contents of this book

Policy- and decision-makers in Ghana will require science-based knowledge to inform the development and implementation of a comprehensive strategic implementation plan to promote food crops agro-processing. Unfortunately, most research and national agricultural development initiatives in the past have focused mainly on upstream activities such as farming to the neglect of downstream activities such as storage, processing and marketing. There is also a critical dearth of knowledge about the major characteristics of MSMEs in food crops agro-processing. It is in the light of the demand for the vital information required for a comprehensive national strategy and the significant knowledge gap that needs to be addressed that this book was conceived. Because it is a demand-led knowledge product, it is more oriented towards addressing policy- and decision-making issues and less towards meeting rigorous academic research requirements. It is only Chapter 4 that is on concentration theories and Chapter 7 is on production and efficiency functions focusing on in-depth analysis of economic theories, estimation and discussion of model parameters.

The book provides science-based information that will help policy- and decision makers to address questions such as:

- *What is the geographical distribution and sector concentration of firms involved in food crops agro-processing in Ghana?*

- *What elements should guide for example the decision to consolidate micro-scale firms to help achieve economies of scale and financial viability of operations?*

- *What are the major factors that determine the efficiency and productivity of the operations of food crops agro-processing firms?*

- *What types and level of inputs such as capital and technology are required to improve the quality, standards and competitiveness of their products?*

  What are the specific characteristics and business conditions of these firms and how do their specific characteristics affect the perception of their informality and financial risks by financial institutions?

- *What are the specific investment and financial needs of MSMEs in agro-processing of food crops in Ghana?*
- *Who are the major financial institutions supplying financial products and services to MSMEs and how adequate and coordinated are these products and services?*
- *What are the key factors behind, for example, the high transaction costs, high business and default risks and what factors determine their access to credit?*

- *What are the investment opportunities available for financing MSMEs in food crops agro-processing and how could financial institutions translate specific financial needs of MSMEs into opportunities for niche financing?*

- *What financial products and services should be structured to address particular financial needs and how innovative and inclusive should these products and services be?*

- *What should policy- and decision makers do to implement appropriate sector strategies, incentives, reforms and mechanisms that will support such innovative and inclusive financial products and services for MSMEs in food crops agro-processing?*

The book is organized into five major parts. Part I includes Chapter 1 and 2 and provides the background, context, objectives and the methodology used. Part II is made up of Chapter 3 to 5 and focuses on the distribution, concentration and ownership structure of MSMEs in food crops agro-processing. Part III includes Chapter 6 to 8 and discusses issues such as food crops agro-processing operations, efficiency and productivity analysis, quality control, standardization, products certification and marketing. Part IV is made up of Chapter 9 to 12 and focuses on issues such as capital investments, analysis of financial needs, access to credit, demand and supply of financial products and services, and financial ratios analysis. Part V is made up of Chapter 13 and discusses some of the critical strategies for intervention and national implementation plan for promoting food crops agro-processing in Ghana.

CHAPTER TWO: GENERAL APPROACH AND METHODOLOGY

2.1. Introduction

Most of the estimation and analytical methods such as firm concentration ratios, production and efficiency functions are discussed in the relevant chapters. This chapter gives a brief discussion of the methodological approaches used mainly for data collection. The study used both primary and secondary sources of data. Secondary sources include review of articles, books, official documents and reports. Chapter three is wholly based on review of relevant secondary materials on the agro-processing sector in Ghana. The primary sources of data used are from questionnaire survey, field observations as well as from formal and informal interviews.

2.2. Theoretical Concept

The theoretical basis of this study is the economic theory of the firm. The theory of the firm consists of a number of economic theories such as the transaction cost theory, production theory and the managerial and behavioral theories that describe and predict the nature of the firm, existence, behavior, structure, and relationship to the market including consumers (Coase, 1937; Baumol, 1962; Williamson, 1966). The theory answers some basic questions such as why firms emerge and what explains their concentration? Why firm owners choose to provide specific goods and services and at particular scale, quality and quantity? What determines the sizes and structure of the firms and what underlies the decision-making, competitiveness, performance and growth of the firm?

Food crop agro-processing firms operate under the competitive market situation. Their establishment, choice of product and operating capacity as well as their efficiency and productivity are determined by several socio-economic factors. Consistent with modern theory of the firm, agro-processing firms exist based on short run motivation of profit maximization and long run motivation of sustainability. They also exist to maximize shareholders profit whiles providing utility to consumers. The study utilizes some of these economic concepts to provide empirical evidence of the operations of food crop agro-processing firms and factors facilitating or constraining their operations in Ghana.

2.3. Data Collection

Field observations, key informant and focus group discussions were the three main qualitative approaches used. For example, the researchers recorded observations on the type and size of assets such as buildings used by the firms. They also noted the scale and level of technology used; product quality and standards as well as measures for quality control of products. Focus group discussions involved mainly micro and small-scale producers organized into cooperatives by MOFA. Issues discussed by the focus group include how they manage production and sales activities such as planning, scheduling, order sourcing and execution of orders; their relationship with suppliers (equipment and other inputs suppliers, service providers, contract farmers; financial institutions) and buyers (credit- receivables, value chain financing). The major constraints and challenges the agro-processing firms are facing and the support (policy, incentives) they will appreciate from the government were also discussed.

Cross sectional data was obtained from agro-processing firm managers and financial institutions using structured questionnaires. Primary data collected included socio-demographic characteristics of firm owners, firm characteristics, capital investment, access to financial products and services, production capacity, operational efficiency and profitability, quality control, sales and marketing. The study focused on firm managers as the key respondents because they have relatively better knowledge of the operation of the firm and are more equipped and disposed to provide information during the interview. However, for most of the micro and small-scale firms in particular, the managers interviewed were also firm owners.

2.4. Sampling Method

In order to have the diversity of food crops processed as well as the diversity of size, scale and scope of the agro-processing firms, Ghana was stratified into major producing regions for cereal (maize, millet, sorghum, and rice) tubers and starchy staples (cassava, yam and plantain), legumes (groundnuts, cowpea and soybean) and fruits and vegetables. The regions were then categorized into major food crops produced based on data from MOFA (2012) on estimated metric tons of production. Due to the predominance of regions in the tubers categories, Brong Ahafo and Western Regions were randomly selected. Eastern

Region was selected for fruits and vegetables, Central Region for cereals and Northern Region for legumes. Figure 2.1 shows the regions selected for the study.

Figure 2.1: Map of Ghana Showing the Selected Regions for the Study.

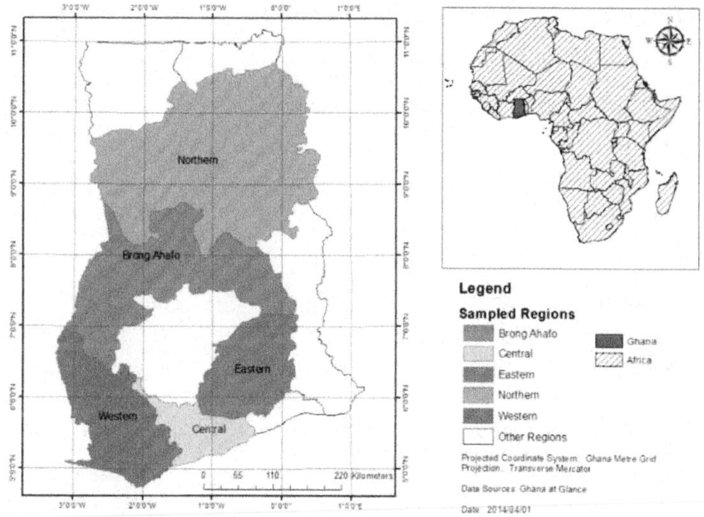

Credit: Kwabena O. Asubonteng

Data on agro-processing firms from MOFA (2010) was updated with information from MOFA regional and District Directors, District Chief Executives and officials of the Ministry of Local Government and Rural Development (MLGRD) and National Board for Small Scale Industries (NBSSI). The target respondents for the study are food crops agro-processing firms, so in order to reduce the costs of the survey, districts with agro-processing firms were isolated and grouped together for each region. The case districts were randomly selected from this group. Firms were then stratified into very-small, small, micro and medium based on the firm sizes. With assistance from the MOFA District Directors and other officials, the firms were randomly selected from the size categories. Overall, 272 firms were randomly sampled for the study.

Table 2.1 provides a summary of the proportion of firms sampled from the various regions. The low number of respondents from Eastern Region was due to the predominance of medium to large-scale oil palm and fruit processing firms in this region and the reluctance of management to allow inquiries into their business operations to allegedly safeguard their competitiveness. The information obtained from most of these firms was therefore not useful for analysis.

Table 2.1: Frequency Distribution of Firms Sampled from the Regions

Major Crops	Region	Frequency	Percent
Roots and Tubers	Western	96	35.3
	Brong Ahafo	60	22.1
Fruits and Vegetables	Eastern	12	4.4
Cereal	Central	63	23.2
Legumes	Northern	41	15.1
	Total	272	100.0

Source: Field Survey 2012

2.5. Limitations of the Study (Challenges with data collection)

Throughout the study, measures were put in place so that the research findings accurately reflect the situation. That notwithstanding, there were some limitations worth mentioning. There were issues of language barrier which slowed down work. One of the main researchers is not a Ghanaian so have to speak through an interpreter and also record the interpreted responses. There is a possibility of miscommunication and misinterpretation of information during this triangular dialogue. Another issue was data fatigue. Some firm managers claimed they have been interviewed for several studies but have not yet seen any benefit of such data collection. However, participation in the survey was voluntary and respondents were made to understand the objectives of the study.

Was the measurement error so insignificant that repeated research with the same empirical indicators and methodology may generate equivalent results? To address this question, the researchers chose a system of measurement and coding consistent with acceptable research guidelines. It was observed during the field study that standards of measurement used by firms were considerably and input measurement was by 'instinct'. There are no standard procedures and recipes with defined amount of inputs for specific unit of processed output. Quantifying and costing the inputs with standardized measurement was a huge challenge.

The study therefore relied on traditional measurements such as "olonka', 'sack,' 'kotoku' for weights and cup or 'kopoo', 'gallon', and bucket for volume during the interview. The figures provided were converted using related standardized units of weights and volumes. Most of the firms were not also keeping data and thus a re-call of past transactions was poor. However, efforts were made to confirm the information from appropriate line managers of the firm. It is appropriate to acknowledge that, the subjectivity involved in the measurement, costing and pricing of inputs and outputs could make the study results vulnerable to biased judgments. This may affect the consistency of the results in multiple surveys.

However, "the measurement of any phenomena always contains a certain amount of chance error. The goal of error-free measurements – while laudable –is never attained in any area of scientific investigation" (Carmines & Zeller, 1979, p. 11). Moreover, "the discrepancies between two sets of measurements may be expressed in miles and, in other cases, in millionths of a millimetre, but if the unit of measurement is fine enough in relation to the accuracy of the measurements, discrepancies will always appear" (Stanley, 1971, p. 356). Chance error in measurement is therefore expected but the degree of error is something that is crucial to the reliability of the research results and its usefulness. The researchers took great care to ensure that the degree of error in measurement is minimized.

CHAPTER THREE: THE FOOD CROPS AGRO-PROCESSING SECTOR IN GHANA

3.1. Introduction

Ghana's agricultural sector is sub-divided into Crops, Cocoa, Livestock, Forestry and Fisheries. The crop sub-sector which contributes about 66.2 percent to the sector has most of its products undergoing primary processing. The processed outputs are categorized into confectionery products, cocoa products, fish and sea foods, horticultural (fruit and vegetables), oil and cereals products (MoFA, 2010). The major raw materials that are involved in these processing include cocoa, cashew, sunflower, oil palm, groundnut, soybean, cotton, cassava, sweet potato, yam, fruits and vegetables.

The Ghana Living Standard Survey (GLSS) report indicates that the major item processed in Ghana is maize. About 283,000 households are involved in processing maize into flour. Tens of thousands of households are also involved in the processing of nuts and pulses into edible oil, fish processing and processing of cassava into flour, dough and gari. There is also the processing of other grains such as millet, sorghum and guinea corn into flour and into local drinks such as 'pito', 'brukutu' and 'asana'. There is also the processing of groundnut into paste. Significant proportions of the processed foods by households are traded in the local markets. For example, about 92.2 percent of the households involved in processing rice sell at their local markets for income. The total sales of home processed maize flour in 2010 amounted to about GHC320,000 (GHC 1= US$0.54) and that of home processed fish was GHC290,000 (GLSS 2008).

3.2. Processing of horticultural products (fruits and beverage)

Ghana's horticultural sector has been very important especially in its contribution to exports. The horticultural products that are processed in Ghana include pineapples, papaya, mangoes, coconuts and fruit. The final products are mainly in the form of fruit salad and juice. Some few medium scale companies have dominated the processing of fruits for exports though there are several micro firms processing fruits for local consumption. For example, Blue Skies of the United Kingdom process some 15,000 tonnes of pineapples at Nsawam for export as fresh cut product. Also, Pinora, a German-Ghanaian firm based in Asamankese buys pineapples and oranges from about 25,000 farmers and processes them into frozen juice concentrates for exports. This means, there are lots of potentials for export earnings, accruable to investors. Other firms involved in these commercial crops are Koranco Farms, Combined Farms, Greentex and Green Span (MLGRD, 2013).

Palm wine is another major beverage produced in Ghana apart from fruit juice discussed above. According to Chandrasekhar et. al. (2012), palm wine is the fermented sap of various palm trees especially Palmyra, Silver date palm and coconut palms. Palm wine also known as 'nsafufuo' in Ghana (Akan Language) is produced by felling the palm tree, boring a hole into the top of the trunk and tapping the sap. The Sap from the palm is a cloudy whitish beverage with a sweet alcoholic taste and with very short shelf life. The palm sap is further processed by fermentation into an alcoholic beverage.

Quote 3.1

Vegetable farmers sign pact to market produce

Thirty-nine vegetable Farmer-Based Organisations (FBOs) in the Greater Accra and Volta regions have signed contracts with various trade associations to supply their produce to them. . With the signing of the contracts, the FBOs will now have ready markets for their produce.
GNA, 2013

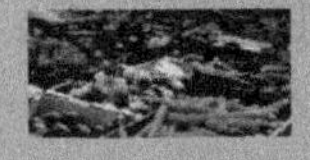

3.3. Processing of Vegetables

Processing of vegetables such as okra, tomatoes, pepper, onions, cabbage, lettuce, green beans and mushrooms are very limited in the country. Although almost all farmers cultivate these crops in their farms, they are usually on subsistence basis for household use. Few sell vegetables from their farms. There are some small-scale vegetable farmers usually at the urban centers who supply customers looking for fresh vegetables. Due to the desirability of fresh vegetables, demand for processed vegetable is very low in Ghana. Those who use processed vegetables such as canned tomatoes usually buy cheap imported products.

This is one of the major reasons why processing of tomatoes which is one of the major vegetables grown at useful scale has not been successful in Ghana. Very few firms therefore go into the processing of vegetables. The survival rate of vegetable firms is very low. As illustrated in Quote 3.1, greater percentage of farmers and firms involved in vegetable farming sell their produce as "fresh" to customers. Without ready market to absolve the vegetable produce, greater percentage is lost through spoilage. The high perishability of vegetables such as tomatoes and the absence of firms processing vegetables at appropriate scale underlie the high post-harvest losses of vegetables – especially tomatoes in the country.

3.4. Processing of Cereal

The two major cereals processed are maize and rice. Other cereals like sorghum and millet are also processed. Maize is usually milled into flour and used for porridge, beverages and other foods. Premium Foods is one of the leading processors of maize and soya beans in Ghana, with an annual production volume of about 10,000 metric tons. Rice processing in Ghana is normally done in the Volta, Eastern and Northern Regions. Processing of rice in Ghana is characterized into household processing, small-medium scale custom milling and large-scale market oriented rice processing based on the scale of operation.

Berisavljevic (2003) notes farmers in Ghana usually thresh the paddy immediately after harvest and store it in bags or on the panicle in special storage facilities. The threshing is usually done on bare floors, under direct sunshine, and also on the road side which result in soil particles and stones getting mixed with the paddy and consequently leading to low-grade rice. In Ghana many villages de-husk rice grain by pounding with a pestle and mortar. In the Northern Region of Ghana, where the climate is much drier, the rice grown in this environment is subjected to the process of parboiling to prevent breakage during milling. The parboiled rice is milled and then sold to the nearby markets.

3.5. Processing of Roots and Tubers

Roots and tubers contribute about 46 percent to Ghana's agricultural GDP (Stumpf 1998). The RTIMP is the program that coordinates the processing of roots and tubers in Ghana. Its task is to identify and promote more efficient processing technologies for roots and tubers. Cassava, the most processed crop among the root and tubers in Ghana have many firms involved in it. Processing of cassava into shelf-stable products and making it fortified with high protein food such as soy helps overcome the high perishability and low level of proteins in cassava. Cassava is processed into a variety of products such as 'gari', 'kokonte', starch, dough, 'tapioca', chips and flour.

Figure 3.1. Some Women Peeling Cassava for Processing in Eastern Region

Source: Field Survey 2012

3.6. Processing of Oil Palm

Palm oil constitutes the single most important source of edible oil for most West African countries especially those in the coastal and forest zones. It is usually used for margarines, cooking fats and soap production. According to Ata (1974), Ghana processes oil palm into two forms namely the 'Dzomi' and 'Amidze'. In order to make the Dzomi, the fresh palm fruit is washed, boiled and pounded in large mortars using pestles to ease the mesocarp from the nuts. The nuts are then removed and the fibrous mash is kneaded to extract the oil. Cold water is then poured into the mash and after some period the oil phase settles on top of the water phase containing the extract from the mesocarp. After the settling period, the oil is skimmed off and retained for further boiling and a little amount of salt is added. The boiling is continued until effervescence ceases. The oil is then left to stand until the roasted fibrous portion of the palm fruit settles at the bottom and separated from the Dzomi. The Amidze oil processing differs from the Dzomi in that there is no boiling of the fruits and the palm fruit used is usually fermented (Ata, 1974). Currently Ghana Nuts is one of the leading agro processing firm involved in the production and export of a range of edible oils; animal feed input materials and shea butter.

Figure 3.2. A man milling palm nuts

Source: Field Survey 2012

3.7. Constraints facing the Agro-processing sector

The agro-processing sector in Ghana is confronted with several challenges. Most important of them is the lack of access to appropriate technology. Most agro-industries in Ghana operate using indigenous technologies. This is as a result of the high cost of processing equipment and the limited capacity of firms to mobilise capital to purchase labor saving technologies especially when it has to be imported. This makes agro-processing operations time consuming, labor intensive with limited opportunity for scaling up. The reliance on rudimentary technology reduces efficiency and the productivity of operations. It also poses some risks to the health of the workers in the firm. Inadequate financing and Credit is also a long standing problem affecting the development of the small scale industries.

There is also the problem of limited access to international markets. In fact, marketing of small-scale processed food product is found to be largely informal. Derbile et. al. (2012) notes the agro-processing enterprises located in the rural areas rely on demand from the small, local, informal and unreliable market. The low quality goods with poor packaging are not able to compete in the international markets with high quality standards.

Most of these firms are also confronted with several safety and health risks. In investigating ergonomic issues in the Ghanaian agro-processing sector, McNeill (2005) used a variety of Participatory Rural Appraisal techniques and a modified hazardous ergonomics risk checklist in seven common post-harvest agricultural systems in Ghana. The various processing firms investigated are cashew nut processing, 'gari' processing, groundnut oil processing, palm kernel oil processing, 'pito' brewing, soap processing and milling. The results of McNeill (2005) survey reveal that 70% of all tasks undertaken in agro-processing were considered to involve repetitive motions, the cause of trauma disorders. Also 55.4% of all agro-processing tasks were undertaken in odoriferous, dusty and smoky environments, which is also detrimental to health of agro processors.

There are also the challenges of lack of managerial skills and training, inadequate finance.

3.8. Some interventions in Ghana's Agro-processing Sector

Over the years, the Government of Ghana has implemented many policies and projects with the focus on improving value addition to Ghana's raw agricultural products through agro-processing industry. The study briefly discusses some of these policies and their relevance for improving agro-processing of food crops in the nation.

The medium term agriculture development program which was adopted by Ghana in 1995, sought to increase productivity in the acquisition and distribution of appropriate inputs and improve agro-processing. This program is Ghana's first move to improving agro-processing after realizing that the agro-processing sector was operating under low efficiency. After having identified inadequate processing facilities and low investment into the agro processing sector, MoFA, through the ***Food and Agriculture Sector Development Policy*** (FASDEP) in 2002, pursued improving processing of raw materials for sustained agricultural development and growth. The Growth and Poverty Reduction Strategy (GPRS II) adopted in 2005 emphasized on promoting agro-processing as part of the efforts to improve private sector development. It proposed several measures to ensure that the agro industry is developing.

The GPRS II promoted the processing and preservation of crop, animals and fish products. The use of standardized packaging materials in addition to efficient application of weights and measures were some of the targets of GPRS II for the agro processing industry (NDPC, 2005). The strategy was premised on the notion that, for the agricultural sector to lead the economy, there should be the necessary inputs for a vibrant agro-processing industrial sector. Therefore, in order to ensure food security for all and increase the access of the poor to adequate food and nutrition, GPRS II sought to facilitate the establishment of small-scale agro-processing industries for export, promote the establishment of storage facilities and develop user - and environmentally-friendly technology for food processing.

Ghana's current agricultural plan, the Medium Term Agricultural Sector Plan (METASIP – 2011 to 2015) which was developed based on the objectives of the second phase of the FASDEP also emphasizes on improving agro-processing to achieving increased agricultural sector contribution.

The plan notes that there are hardly any statistics on processed agricultural products in the country; however, value addition to primary products is a critical element of agricultural modernization. The METASIP really has some promising highlights for the growing agro processing industry. It seeks to promote the use of storage facilities such as silos, cold storage units and modern markets in strategic locations to minimize post-harvest losses.

It also seeks to enhance returns for small scale farmers through equitable access to resources and services, promotion of high value crops and high quality processed products. It places priority on the processing of major food staples such as maize (milling and packaging), rice (milling and packaging), cassava (gari, flour, etc), yam (flour) and cowpea (grading and packaging). Consistent with the provisions of the GPRS II, METASIP emphasizes the use of appropriate grades and standards to improve quality, market penetration and competitiveness of products from Ghana. Other activities highlighted in the METASIP to improve the food processing ventures include assessing the quality of agro-processing technologies, developing standards for agro-processing equipment for various types of food products, enforcing the use of food grade equipment and facilitating access to credit facilities.

The Root and Tuber Improvement and Marketing Program (RTIMP) is being implemented to address the issues relating to processing of root and tubers in Ghana. The RTIMP in collaborations with District/Municipal Agriculture Development units seek to identify strategic processing firms to upgrade them to Good practices Centres to serve as important centres of the cassava value chain.

The RTIMP centres will also be equipped with the capacity to absorb raw cassava from farmers particularly during the harvesting season. The Ghana Commercial Agricultural Project (GCAP) also seeks to improve agro-processing in horticultural sector in order to provide alternative markets to local production. The Savannah Accelerated Development Authority in partnership with the Government has also established three Agro processing factories- a Sheanut processing factory at Buipe, a Rice mill at Nyankpala near Tamale and a Vegetable oil mill at Tamale (MoFA 2013).

In a new alliance for food security and nutrition in Ghana, the government of Ghana in collaboration with the G8 Cooperation are working together to increase private sector investment in agricultural development, to scale up innovation, and achieve sustainable food security outcomes. These initiatives are intended among other objectives to improve rural income, employment, poverty reduction and food security. Several private companies have expressed interest in investing in Ghana's agro processing. For instance, Ecobank Ghana has targeted lending US $5 million to agribusiness small-and medium-sized enterprises operating in the rice, maize, and soya value chains. The Ghana nuts have a plan to increase the procurement of soya for processing and create a reliable market for soya farmers.

3.9. Conclusion

The agro-processing sector is currently plagued with technical, commercial, financial and managerial challenges. Since 1995, the promotion and growth of agro-processing have been considered in several policies and programs by the government. However, to date, the achievement of the expected policy impacts and outcomes has been very modest. One of the reasons is the lack of comprehensive and strategic implementation of policies and activities focusing on critical areas such as access to finance, industrial infrastructure, technology and incentives (including tax rebates) to agro-processing firms. The agro-processing sector could be one of the main avenues for industrialization, value addition, employment and inclusive growth in Ghana. The Government and private institutions have realized the investment potential and the opportunities for improving this sector and this presents timely opportunity for a comprehensive national policy for agribusiness and agro-industries development in Ghana. Certainly, there is the need for new knowledge that will help support strategic implementation of activities to build on some of the past initiatives.

One of the major objectives of this book is to provide information that will help policy-and decision makers to develop comprehensive and strategic plans to scale up agro-processing ventures in the nation.

CHAPTER FOUR: DISTRIBUTION AND CONCENTRATION OF FOOD CROPS AGRO-PROCESSING FIRMS IN GHANA

4.1 Introduction

This chapter focuses on the measurement of concentration and distribution of food crops agro-processing firms in Ghana. This is intended to provide information on the extent of competition and concentration in the food crop agro-processing sector. Information from these analyses may be useful for business policy and further development of the sector.

Food crops agro-processing firms are facing increasing external competition due to dumping of cheap price and sometimes relatively low quality foreign products. The prevailing competitive market environment may be one of the justifications for the promotion of medium to large-scale firms with the capacity to compete in the processed food market. However, this may lead to higher concentration of food crops agro-processing firms and possibly to a monopolistic system. Besides, small-scale food crop agro- processing activities represent a potential source of livelihood for many poor people in Sub- Saharan Africa including Ghana. For instance, it is estimated that 60% of the industrial labor force in Sub-Saharan Africa are employed in small-scale food processing enterprises and the majority are women (ITDG, 2005). The loss of business by micro- and small-scale processors may lead to massive retrenchments, and worsen the already high unemployment and underemployment particularly in rural areas. As indicated above, this will have relatively greater impacts on women - majority of them being firm-owners or employees.

In order to have effective policies that will enhance the competitiveness of food crops agro-processing firms without adverse effect on micro- and small-scale firms, it will be necessary to understand the distribution and concentration of food crops agro-processing firms throughout the country. Economic theory suggests that concentration is an important determinant of market behavior. It is also important to understand some of the linkages between localization and sizes of firms, competitive advantage, types of products and scale of operation, market share and nature of competition among the firms.

4.2 Types and Classification of Agro-processing Firms

Agro-processing firms have generally been classified based on the level of processing or value addition, the type of products produced, level of upstream or downstream processing or product use and the size of the firm. Food crop agro-processing firms include firms involved in processing of fruits and vegetables, oil products, cereal, roots and tubers and other processed food products. There are other agro-processing firms involved in the processing of meat, fish, and the production of cloths, foot wares and herbal medicines. However, these agro-processing firms are outside the focus of this study.

Based on the level of processing, food crops agro-processing firms can be classified as primary or secondary firms. *Primary processing firms* are involved in activities that are mainly carried out at the farm and may not transform or slightly transform the commodity into different form prior to storage, marketing or further processing. These firms usually use simple manual tools in their processing operations. Some of the activities undertaken by these firms include, shelling/threshing, crop drying, cleaning, grading, and packaging. *Secondary processing firms* increase the nutritional or market value of the commodity by changing the physical form or appearance or the chemical composition of the commodity. Secondary processing operations therefore may involve milling or grinding, pressing, mixing, heating and others.

Another commonly used method for categorizing agro-processing firms is the size of the firm defined in terms of total volume of output, gross income or sales, number of employees and total company assets. Based on this criterion, agro-processing firms are grouped into *micro, very small or small, medium or large scale enterprises.* But the issue of what constitutes a micro, very small or small, medium or large scale enterprise remains a major concern in literature. Whiles some uses the capital assets and/or turnover, others rely on the number of employees (Storey, 1994; Abor and Quartey, 2010). Even within the same country, variations in the definition of the size of the company may exist.

According to the Bolton (1971), a firm may be classified from an economic and statistical point of view. Under the economic definition, the Committee uses relative share of the market (market share); management structure of the firm and level of dependence on labor and technology as the main criteria for grouping firms as micro, small or medium enterprise. The statistical definition considers the firm's contribution to gross domestic product, employment, exports, and how its economic contribution to the particular sector has changed over time.

Yaron (1997) indicated that small and medium-sized enterprises are defined according to their staff headcount and turnover or annual balance-sheet total in the European Union (EU). The EU categorizes companies with less than ten employees and annual turnover and/or annual balance sheet ≤ EUR 2 million as micro; those with less than 50 employees and annual turnover and/or annual balance > EUR 2 million and ≤ EUR 10 million as small firms. Firms with less than 250 employees and annual turnover > EUR 10 million and ≤ EUR 50 million are classified as medium-scale (Yaron, 1997). By contrast, in the United States and Canada, small firms often refers to those with less than 100 employees, while medium-sized firms often refers to those with less than 500 employees (Carsamer, 2009).

There are several firm classifications in Ghana based on different criteria. For example, Kayanula and Quartey (2000) classify firms into micro, very-small or small and medium scale according to the number of employees of the enterprise. The firm classification by the Ministry of Trade and Industry (MOTI) consider businesses that employ up to five employees with fixed assets (excluding real estate) not exceeding the value of $10,000 as micro enterprises (Mensah, 2005). Small enterprises are those that employ six to twenty nine employees with fixed assets of $100,000 and medium enterprises are business entities that employ thirty to ninety-nine employees with fixed assets of up to $1 million.

According to the National Board for Small Scale Industries (NBSSI, 1990), a micro-enterprise is a firm with employee less than five and a small-scale enterprise is a firm with not more than nine workers, and has plant and machinery (excluding land, buildings and vehicles) not exceeding ten million Ghanaian Cedis. The Ghana Statistical Service (GSS) categorizes firms with less than ten employees as small-scale enterprises and those with more than ten employees as medium-sized enterprises (Kayanula and Quartey, 2000). According to Osei et al (1993) a micro enterprise is a firm employing less than six people, a very small enterprise is that employing six to nine people; and a small enterprise is that with between nine and thirty employees.

The foregoing discussions therefore indicate the complexity in definition, classification, and methods with varying cut off points making it increasingly difficult to draw a clear distinction between what firms should be considered as micro, very small, small, medium or large scale. This study used operational definition based on the number of employees of the enterprise. It considers a micro-scale agro-processing firm as an entity that employs less than six people, a very small-scale agro-processing firm as that employing six to nine people, a small-scale agro-processing firm as those that employ ten to twenty nine people, and a medium-scale agro-processing firm as those employing between thirty and hundred people.

As illustrated in Figure 4.1. 85% of agro-processing firms in Ghana are micro-enterprises. This finding supports NBSSI (1991) findings that about 90% of firms in Ghana are micro or small scale enterprises. Very small firms which employ about six to nine people constitute 7%. Small agro-processing firms form only 5% whereas medium agro-processing firms form 3%.

Figure 4.1 Size Distribution of Agro-processing Firms in Ghana

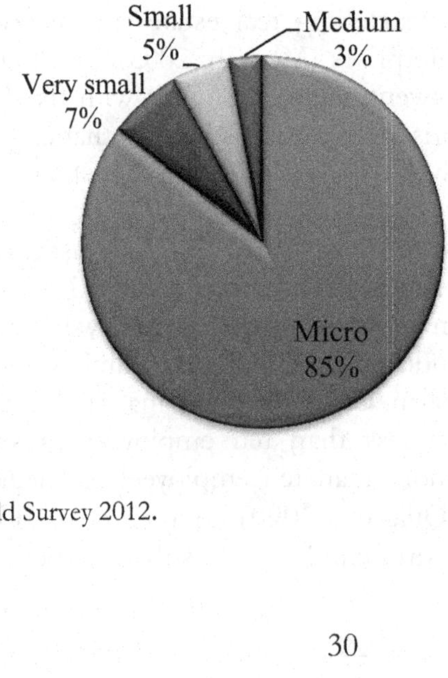

Source: Field Survey 2012.

4.3 Distribution of Food Crop Agro-Processing Firms in Ghana

In describing the distribution of agro-processing firms in Ghana, the firms have been clustered into various forms based on some common parameters such as major commodity produced. Frequencies and percentages are used to describe firm features such as ownership characteristics, the type of commodity produced, and distribution of different sized agro-processing firms across the study regions.

4.3.1 Types of Products Processed by Agro-processing Firms

An agro-processing firm may be involved in the processing of multiple products. Therefore multiple responses are possible in terms of types of products produced. However, a firm may have a key or major commodity it produces. Other products may be subsidiary or supplementary products. These major products are often determined by the firm, based on gross returns from sale of that product or net benefit from the product. From Table 4.1, agro-processing firms that produce oil and oil by-products, and roots and tuber related products constitute the largest group, accounting for 56.6% and 24.3% respectively of the surveyed sample. Cereals, and soap and cosmetic producing agro-processing firms account for 8.8 and 5.5 per cent respectively. Only 2.6% and 1.5% of the agro-processing firms are into production of alcoholic and non-alcoholic beverages such as fruit juice. As noted in Section 2.4, the problems encountered with securing useful information from several of the medium-scale fruit processing firms during the field study in the Eastern Region may have contributed to the low percentage of non-alcoholic beverage firms.

Table 4.1 Distribution of Agro-processing Firms Based on Type of Product Processed

Product type	Frequency	Percent
Roots and tubers	66	24.3
Cereals	24	8.8
Vegetables	2	.7
Oil and Oil by-products	154	56.6
Soaps and cosmetics	15	5.5
Alcoholic beverages	7	2.6

Non-alcoholic beverages	4	1.5
Total	272	100.0

Source: Field Survey 2012.

The results also show that agro-processing in Ghana are skewed towards two major product categories- oil, and root and tuber processing. As illustrated in Table 4.2, palm oil producing agro-processing firms account for more than half of all agro-processing firms in Ghana. The skewness of agro-processing towards particular agro-products may be influenced by demand or availability of market for the products. Also, the Root and Tuber Improvement Programme and the relatively ready internal or international market for roots and tuber products such as 'gari' may be some of the reasons for the high percentage of firms focusing on these products.

Less than 1 percentage of firms interviewed are processing vegetables such as pepper, okra, carrots, and tomatoes. This may be due to the relatively high perishability of vegetables and lack of the financial capacity for the appropriate technology, packaging (including preservatives) and storage facilities needed for viable business in vegetable processing. Unlike other products such as palm oil or 'gari' that may not require substantial capital for storage and specialized packaging facilities, processing of fruits and vegetables require specialized and sophisticated equipment to carry out the operation and management (including storage and marketing) of those products. The initial capital requirement for viable business in vegetable processing may render it very prohibitive to most firms.

Table 4.2 Commodity-specific Distribution of Agro-processing Firms

Commodity	Frequency	Percent
Palm oil	140	51.5
Gari	59	21.7
Rice	23	8.5
Shea butter	14	5.1
Palm kernel oil	10	3.7
Local gin / Akpeteshie	7	2.6
Cassava dough	6	2.2
Groundnut powder	4	1.5

Coconut juice	2	0.7
Fresh fruit concentrate /juice	2	0.7
Corn dough	1	0.4
Okro	1	0.4
Cabbage	1	0.4
Cassava powder	1	0.4
Soap	1	0.4
Total	272	100.0

Source: Field Survey 2012

4.3.3 Sizes of Agro-processing Firms in Ghana and Commodities Produced

As shown in Table 4.3, most of the micro-sized food crop agro-processing firms (61.6%) are into processing of oil and oil related products as well as roots and tubers (22.0%). Processing of root and tuber products dominates among very small-sized food crop agro-processing firms. For small and medium-sized food crop agro-processing firms, processing of soap and cosmetic products as well as root and tubers are the two predominant business activities.

Table 4.3 Distribution of Food Crop Agro-processing Firms according to Commodities Produced

Product type	Size of firm			
	Micro	Very small	Small	Medium
Roots and tubers	22.0	38.9	21.4	62.5
Cereals	9.5	11.1	0.0	0.0
Vegetables	0.9	0.0	0.0	0.0
Oil and oil by-products	61.6	27.8	28.6	25.0
Soaps and cosmetics	3.0	22.2	28.6	0.0
Alcoholic beverages	1.7	0.0	21.4	0.0
Non-alcoholic beverages	1.3	0.0	0.0	12.5

Source: Field Survey 2012

4.3.4 Total Annual Turn-over of Food Crop Agro-processing Firms in Ghana

Table 4.4 presents the distribution of food crop agro-processing firms in Ghana according to their total annual turn-over. Micro-sized agro-processing firms have total annual turn-over between GHC600.00 to GHC88, 000.00. Very small agro-processing firms have total annual-turnover of GHC5, 760.00 to GHC 129,792.00 with a mean income of GHC 39,234.8. Small firms have total annual turn-over in the range of GHC3,888.00 to GHC1,728, 000.00 with a mean total annual income of GHC203,356.8. Medium-sized agro-processing firms on the other hand are those with income range between GHC34, 560 to GHC12, 330,603.00, and an average annual turn-over of GHC2, 129,537.4.

Table 4.4 Distribution of Food Crop Agro-processing Firms in Ghana according to Total Annual Turn-over (GHC)

Firm size description	Minimum	Maximum	Mean	Std. Deviation
Micro	600	88000	22073.4	23486.77
Very small	5760	129792	39234.8	40621.48
Small	3888	1728000	203356.8	538774.93
Medium	34560	12330603	2129537.4	4998338.67

Source: Field Survey 2012

The minimum and maximum ranges do not show any clear cut demarcations between micro, very small, small and medium scale agro-processing firm. For example, there are some small-scale agro-processing firms with total annual turnover below the maximum total annual turnover for even micro-scale firms. However, the average total annual turn-over does reflect those distinctions according to the categorization based on number of employment. This wavy demarcation arises due to differences in production and market efficiency. Whiles some micro and very-small firms may be operating at full capacity, some small and medium-sized firms may be operating below installed capacity, making their incomes comparable to micro or very small firms.

4.3.5 Regional Distribution of Different-sized Agro-processing Firms in Ghana

The location of an agro-processing firm has influence not only on its production and marketing operations but also on its steady expansion in size. As indicated in Table 4.5, 50% of agro-processing firms interviewed in Eastern region are micro-scale enterprises. However, the Eastern region has the lowest percentage of micro-scale agro-processing firms compared to the other regions under analysis. The Eastern Region also has relatively higher percentage of small and medium-sized agro-processing firms. The rest of the regions have a high percentage of their agro-processing firms being micro-scale enterprises. None of the firms interviewed in the Western Region is a medium-sized food crop agro-processing firm. The absence of many medium-sized food crop agro-processing firms across the regions may be one of the causes of high post-harvest losses, since the micro-sized firms lack the capacity to process all market surpluses, particularly, in times of bumper harvest. Fewer medium-sized agro-processing firms will also hamper downstream agro-based industries.

Table 4.5 Distribution of Agro-processing Firms across Some Regions in Ghana

Firm type	Region					Overall
	Eastern	Central	Northern	Brong Ahafo	Western	
Micro	50.0	92.1	73.2	81.7	92.7	85.3
Very small	8.3	3.2	14.6	8.3	4.2	6.6
Small	25.0	3.2	9.8	3.3	3.1	5.1
Medium	16.7	1.6	2.4	6.7	0.0	2.9

Source: Field Survey 2012

4.4 Measurement of Industry Concentration

The theory of firm and market structures suggests that concentration, defined as the extent to which a small number of firms account for a large proportion of an industry's output, is an important determinant of market behavior. Measures of concentration are also essential for welfare-related public policy. Such policies will be critical for the pursuit of inclusive growth in the country. Moreover, concentration of firms in an industry has been used in business mergers and anti-trust policies in many countries. This therefore has prompted several interest groups and researchers to develop methods for measuring firm concentration.

Variety of approaches have been developed and used in literature to empirically gauge the extent of industry concentration in an economy.

However, the concept of industry concentration can have different connotations depending on what objective the researcher hopes to achieve. Absolute concentration refers to the extent to which *small number* of firms accounts for a large proportion of industry size. Thus, absolute concentration looks at concentration along a single industry. By contrast relative concentration focuses on the extent to which a *small percentage* of firms account for a large proportion of industries size. Relative concentration looks at inequality by comparing concentration across two or more industries.

Methods for measuring concentration include the comprehensive industrial concentration index, Berger-Parker index, Hall-Tideman and the Rosenbluth index, Simpson, Shannon, Ogive, Entropy, Modified Entropy, Composite Entropy, Maurel and Sedillot index, Hannah and Kay index, Davies U index, Hause indices, Herfindahl or Herfindahl-Hirschman index, adjusted Herfindahl index, the Lorenz curve, Gini coefficient, concentration ratio as well as concentration curves. These methods have been used in several areas ranging from transport, banking, marketing and agriculture to measure the degree of concentration, specialization or diversification in an economy (Campos, 2012; IFPRI, 2012; Kelley *et al.,* 1995; Chand, 1996; Pandey and Sharma, 1996; Kacperczyk et al, 2005; World Bank, 2011).

Although these methods mentioned above have their merits and demerits, the best concentration measure is one with the ability to measure important structural features of the industry (Bikker and Kaaf, 2002). Therefore the most widely used in literature include the Berger-Parker index, Shannon index, Simpson index of diversity, the Lorenz curve, Gini coefficient, Entropy and modified entropy, concentration curves and *K*-concentration ratios, and the Herfindahl or Herfindahl-Hirschman index of concentration.

The Berger-Parker index (D) emerged from biodiversity analysis and it is often used to estimate the relative abundance of individual units. It is estimated as:

$$D = \frac{1}{\max(\alpha_i)}; \ D \geq 1 \quad (4.1)$$

Where:α_i denotes the share of the size of ith firm in the industry. The index ranges from$1 \leq D \leq N$, where N is the maximum possible number of firms in the industry. A higher index indicates higher concentration. Davis *et al.*, (2010) used this method to estimate the diversity of incomes from rural income generating activities in a cross country comparative study (Eastern Europe (Albania , Bulgaria); Africa (Ghana , Madagascar , Malawi , Nigeria); Latin America (Guatemala , Ecuador , Nicaragua , Panama); Asia (Bangladesh , Indonesia , Nepal , Pakistan , Tajikistan , Vietnam).

The Shannon Index (SI) is used to assess diversity of species (Magurran, 1988) and is given as:

$$SI = -\sum_{i=0}^{n}[(share_i).ln(share_i)] \quad (4.2)$$

Where: n = total size of firms in an industry, $share_i$= share of *ith* firm size in the industry.

The Simpsons index of diversity (SID) has also been used in many instances to measure the extent of competition among industry players and is estimated as:

$$SID = 1 - \sum_{i=1}^{n} p_i^2 \quad (4.3)$$

Where: P_i = the share of firm *i* in the economy, *n* = number of firms being studied. The index ranges from zero to one, with zero being concentration and one being the extreme value of competition. The Berger-Parker index, the Shannon Index and the Simpsons Index are more suited to measurement of diversification and competition and thus are less preferred in the measurement of industry concentration.

The Lorenz curve and Gini coefficients are also another class of measures of concentration. The Lorenz curve measures the cumulative percentages of output accounted for by various percentages of the number of firms whereas the Gini's concentration ratio, which takes a value between zero and one, is a function of the area between the

Lorenz curve and the diagonal line the curve would follow if all firms were of equal size. Both methods measure inequality or relative concentration (Hagerbaume, 1977).

Absolute concentration is more directly relevant than either inequality or average firm size when the focus of analysis relates to problems of monopoly and business policy (Rosenbluth, 1955). For instance, an index of inequality may tell us that 5 per cent of the firms in an industry control 90 per cent of output, but it does not tell us whether this 5 per cent consists of one firm or perhaps a hundred firms. Certainly, a competitive pattern of behaviour is much more likely in the latter case than in the former. This study relies on traditional measures of absolute concentration such as concentration curves, k-concentration ratios and the Herfindahl-Hirschman index.

4.5. Estimates of the concentration of Food Crops Agro-Processing Firms across Regions in Ghana by the concentration curve

The concentration curve is very simple and a useful approach used to describe the concentration of firms in an industry. It is a graphical presentation of the cumulative distribution of firm numbers and shares on the OX- OY Cartesian axis. The curve continuously rises from left to right, but at a diminishing rate. It starts from the first firm's share and reaches its maximum height at 100 per cent on the OY-axis at which point it corresponds to the total number of firms in the industry when traced to the OX- axis. Thus, the height of the curve above any point x on the horizontal axis measures the percentage of the industry's total size accounted for by the largest x firms. A short, steep curve indicates high concentration while a low-lying, long curve indicates low concentration. The major challenge arises when curves intersect. It is often difficult to say which represents high concentration.

The steepness or flatness of the concentration curve could be used to compare concentration of firms in an economy or across regions. As shown in Figure 4.2, the concentration curves of Western, Central and the Northern regions remain virtually flat and long-stretched. These depict low concentration of food crop agro-processing firms in the regions. Clearly, Western region depicts the least concentration as evident from its low-lying curve. This is followed by the Central and then the Northern region. In the same vein, the concentration curves of Eastern and Brong-Ahafo regions are slightly domed and steep, reflecting relatively higher concentration of firms in the region.

Also, it could be observed that some portions of the concentration curve for Brong-Ahafo region lied below the curves of central and western regions before rising above them. This may illustrate increasing or constant returns of firm sizes to the concentration of food crop agro-processing firms in Brong-Ahafo region as opposed to diminishing returns to concentration of firms in the other two regions. The low-lying portions of the Brong-Ahafo region below the other two regional curves when traced to the horizontal axis also suggest that the concentration of the first largest and the third largest food crop agro-processing firms in the Central and Northern region may be higher than that of the Brong-Ahafo region.

Figure 4.2 Concentration Curves for Food crop Agro-processing Firms across Regions in Ghana

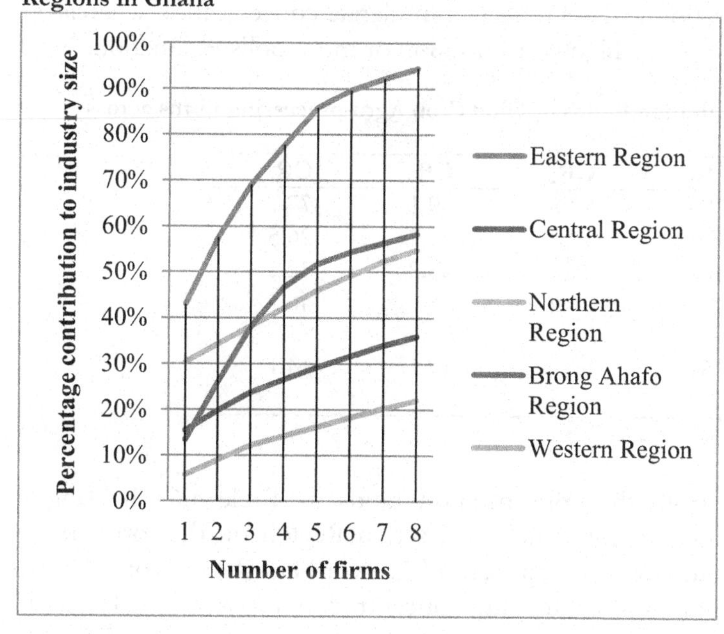

4.6. Estimates of the concentration of Food Crop Agro-Processing Firms across Regions in Ghana by the concentration ratios.

Another widely used method mainly because of its simplicity and limited data requirement is the k-concentration ratios (CR_k). It is defined as the percentage of the industry share owned by the largest k firms in an industry. It is empirically estimated as:

$$CR_k = S_1 + S_2 + \cdots S_k \quad (4.4)$$

Where: S_1 to S_k are firm shares of the largest k firms in the industry. The concentration ratio is often calculated for the largest three, four or eight firms in the industry. For instance, the United States of America uses a four-firm or eight- firm concentration ratio in analyzing their data. The Great Britain however uses three-firm concentration ratio. The centration ratio ranges from zero to hundred. A four-firm concentration ratio of zero to fifty implies a low concentrated industry. A four- firm concentration ratio of 51-80 reflects a moderate concentration whereas a ratio of 81-100 implies high concentration or monopolistic situation.

Table 4.6 Concentration Ratios of Food Crop Agro-processing Firms across Regions in Ghana

Region	CR_1	CR_2	CR_3	CR_4
Eastern	42.9	57.5	69.1	77.7
Central	15.4	19.8	23.7	26.5
Northern	30.4	34.3	38.2	42.1
Brong Ahafo	13.3	25.9	38	46.7
Western	5.8	9	12.1	14.1

Source: Field Survey 2012

As elaborated from the concentration ratios in Table 4.6, the single largest firm concentration ratio put Eastern Region on the top with a concentration ratio of approximately 43%. Northern Region comes next followed by Central Region with concentration ratios of 30.4% and 15.4% respectively. The region with the least concentration is Western Region. Except for few variations in the concentration ratios of Central, Northern and Brong- Ahafo region, the trend in the two, three and four firm concentration is not much different from the single-firm concentration ratio earlier discussed. In the two-firm and four- firm concentration ratios, Brong Ahafo becomes the third and the second most concentrated Region in Ghana.

However, comparing the four-firm concentration ratios of the five regions with the concentration cut-off points proposed by USDOJ (2010), the results points out clearly that most of the regions under study have low concentration of food crop agro-processing firms. Eastern Region is the only moderately concentrated region in Ghana. This may be due to the location of several medium- to large-scale oil palm and fruit juice processing firms in this region. Merger or consolidation of firms in Eastern region will require careful scrutiny and analysis of post-merger effect on concentration. However, most of the food crops agro-processing firms in the other regions can merge or be consolidated without unduly affecting prices and post-merger concentration.

However, concentration ratios present an incomplete picture of the concentration of firms in the industry since it does not use the market shares (competition) of all the firms in the industry. If there is a significant change in the market shares among the firms included in the ratio, the value of the concentration ratio would remain the same. Thus, a common setback in both the concentration curves and ratios is that there are many changes in the position of the firms that leave the index unchanged. The lack of a summary measure utilizing all firm sizes or points on the curve has therefore been criticized and even offered as an argument for using a different concept of concentration (London Economics, 2007).

The Herfindahl index (often referred to as the Herfindahl-Hirschman index (HHI)) offers an alternative measure of industry concentration than the concentration ratio or curve by involving all the firm sizes. Besides involving all firms, it also squares the firm sizes to place weight on the larger firms. Also, the index has an appeal of summarizing a whole vector of levels or shares data in a single number from zero to one. The major strength of the method is that the HHI increases as both the number of firms in the market decreases and also as the disparity in size between the firms increases. A major weakness of the HHI is that two dissimilar industries distribution can give similar values. The HHI is the sum of squares of all proportions of the enterprises. It is expressed in terms of sum of squares of firm sizes, all measured as percentages of total industry size. The HHI can be mathematically specified as:

$$H = \sum_{i=1}^{n} S_i^2 \qquad (4.5)$$

Where: H= Herfindahl Index and $S_{i=}$ proportion of firm size *i* in the total industry and n = total number of firms in the industry. Thus, if there are n firms in an industry, then the Herfindahl index can be empirically estimated as:

$$H = S_1^2 + S_2^2 + S_3^2 + \cdots S_n^2 \qquad (4.6)$$

This index is equal to the reciprocal of the number of firms if all firms are of the same size, and reaches its maximum value of unity when there is only one firm in the industry. The index summarizes shares of data into a single number from zero to one. The HHI is explained in relation to increasing concentration, with extreme concentration (or monopolistic market) assuming the value of one and zero being extreme market competition. Increases in the Herfindahl index generally indicate a decrease in competition and an increase of market power, whereas decreases indicate the opposite. Alternatively, if whole percentages are used, the index ranges from 0 to 10,000. According to the United States Department of Justice (2010), an HHI less than 1,000 implies that the industry is not concentrated; HHI from 1,000-1800 refers to moderate concentration whereas an HHI above 1800 is considered as highly concentrated.

IFPRI (2012) in analyzing economic transformation in Ghana fitted a Herfindahl index for agricultural exports and agro-processing products to measure export diversification. The study found that between 1965 and 2008, the normalized Herfindahl index for Ghana fell from 50–60 before the mid-1980s to 28–38 in recent years, suggesting that export diversification has occurred. The study concluded that Ghana has not experienced the kind of agricultural productivity growth that can drive the development of competitive agro-processing industries and that difficulties faced by agro-processing in the country suggest that more competitive agriculture in this country can contribute more to economic transformation.

As shown in Table 4.7 and based on the USDOJ (2010) classification of firm concentration using the HHI, Western, Central, Northern and Brong-Ahafo regions are not concentrated with food crop agro-processing firms. Consistent with results of the concentration ratio and curve, Eastern region is however, moderately concentrated with food crop agro-processing firms.

Table 4.7 Herfindahl-Hirschman Index for Food Crop Agro-processing Firms across Regions in Ghana

Region	N	HHI
Eastern	12	2364.9
Central	63	398.5
Northern	41	1079.7
Brong Ahafo	60	643.1
Western Region	96	157.9

Source: Field Survey 2012

Comparing the result from the HHI of the five regions in Ghana using analysis of variance shows that there are significant differences among the regions in terms of the concentration. Whereas Eastern region is significantly more concentrated with food crops agro-processing firms than the other regions in Ghana, Northern region comes second with Western region being the least.

Table 4.8 Analysis of Variance Table for Concentration of Food Crop Agro-processing Firms across the Regions

Region	Mean	F	Sig.
Eastern	197.1		
Northern	26.3		
Brong Ahafo	10.7	7.162	0.000
Central	6.3		
Western Region	1.6		

Source: Field Survey 2012

As shown from the mean plot, the results also show wide margin between the highest concentrated region- Eastern region- and the rest of the regions. There is also a sharp fall or decline from the Eastern regional centroid to the rest of the regions. Whiles Eastern region lies up separated from the rest of the regions, the remaining four regions lie nearly on a plateau.

Figure 4.3 Mean Plot of Agro-processing Firms across the Regions

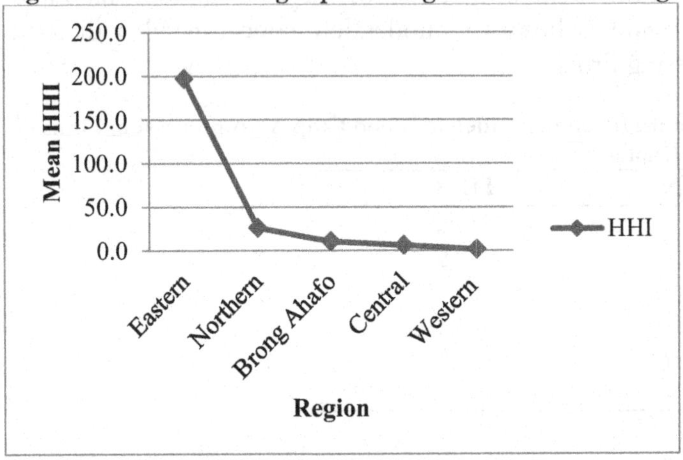

Source: Field Survey 2012

4.7 Concentration of Food Crop Agro-Processing Firms across Commodity-Types in Ghana.

When concentrations of firms are looked at in isolation from the products the various firms are involved in, there is a potential danger of overlooking the significant relationships between concentration and competition in the economy as well as the policy implications. Thus, in studying the concentration of firms in an industry, not only is it appropriate to look at the geographical concentration but also the commodity-specific concentration. Based on our classification of commodities produced by the food crop agro-processing firms in Ghana, non-alcoholic beverages (fruit juice), oil and oil products sub-sector have the highest concentration as shown in Table 4.8. As observed earlier in section 4.6, most of the firms in these sub-sectors are located in the Eastern Region. Other things remaining constant, mergers of agro-processing firms in this sub-sector of the industry may have negative consequences on consumer welfare especially if it involves the largest firms in this product category. Intense competition between producers of vegetable, cereal and alcoholic beverages is possible considering their low HHI. To benefit from economies of scale, it would be appropriate for firms in these sub-sectors to merge or be consolidated since concentration is very low.

Table 4.9 Product Concentration of Food Crop Agro-Processing Firms in Ghana

Product type	N	HHI
Roots and tubers	66	997.48
Cereals	24	46.04
Vegetables	2	.31
Oil and Oil by-products	154	1597.11
Soaps and cosmetics	15	106.91
Alcoholic beverages	7	52.40
Non-alcoholic beverages	4	1843.90

Source: Field Survey 2012

Comparing the result from the HHI of the based on the product categories using analysis of variance shows significant differences between the products in terms of concentration (Table 4.10). The results further indicate great disparity in the level of concentration of firms in the specific product types with the highest significant concentration observed in the non-alcoholic beverage sector.

Table 4.10 Analysis of Variance Table for Concentration of Food Crop Agro-processing Firms across the Regions

Product	Mean HHI	F	Sig.
Non-alcoholic beverages	460.9747	9.910	0.000
Roots and tubers	15.1133		
Oil and Oil by-products	10.3708		
Alcoholic beverages	7.4864		
Soaps and cosmetics	7.1275		
Cereals	1.9182		
Vegetables	0.1562		

Source: Field Survey 2012

4.8. Conclusion

The study has shown that majority of the food crops agro-processing firms in Ghana are micro-enterprises involved in processing of oil products. Fewer firms are into fruits and vegetable processing. Research should therefore be geared towards identifying the limiting factors that hinders small scale food crop agro-processing firms to go into fruits and vegetables processing. Similarly, studies should also look at micro-enterprises, whether it has any strength or delimiting factors that initiate growth into medium or large scale food crop agro-processing firm.

Generally, firms are not concentrated across the regions. This has implications on unfair competition considering the size of the Ghanaian economy. Eastern region remains moderately concentrated. This may have implications on market structure and results in the region. This should be monitored carefully. Since there are no or limited barriers to internal trade in Ghana, the high concentration resulting from failure to measure and control concentration could have a spill-over effect even on nearby regions or the country as a whole. Concentration is also high in the non-alcoholic beverage and oil-processing firms, yet many more micro-enterprises are found in this sector. This demands further research to unravel the mystery behind this trend.

CHAPTER 5: OWNERSHIP AND EMPLOYMENT CHARACTERISTICS OF FOOD CROPS AGRO-PROCESSING FIRMS IN GHANA.

5.1. Introduction

Ownership of agro-processing firms may provide significant information that could guide policy makers in taking relevant decisions pertaining to the sector. For example, in order to design and implement the appropriate gender-based policies for promoting agro-processing firms in Ghana, it will be worth knowing the ownership and firm characteristics by gender. Firm-owner characteristics include demographic, individual and personal traits. The firm characteristics refer to the origin of enterprise, length of time in operation, size of enterprise and sources of capital. These firm characteristics may provide important information in determining success or failure of the business. The individual characteristics of firm owners discussed in this chapter include attributes like age, sex, education, managerial know-how, industry experience and social skills of the owner. The length of time deals with the duration the firms have been in operation whilst the size of enterprise deals with how large a firm is with respect to employment (Islam et. al, 2011).

Several studies have indicated that firm and ownership characteristics have their respective effects on certain indicators of the success of businesses. For example, Cacciolatti and Wan (2012) assessed small business owners' personal characteristics and the use of marketing information in the food and drink industry. They collected data through a regional survey of 296 small business owners and senior managers in SMEs in the Scottish food and drink industry. The study used a canonical correlation analysis and found that personal characteristics such as age, gender, previous experience, and marketing expertise are critical factors affecting information use among business owners. Barbieri and Mshenga (2008) investigate the role of firm and owner characteristics on the gross income of farms engaged in agri-tourism[1].

---

[1] Any activity in which a visitor to the farm or other agricultural setting contemplates the farm landscape or participates in an agricultural process for recreation or leisure purposes

Their findings indicate that the length of time in business, the number of employees and the farm acreage have a positive impact on performance in terms of annual gross sales of agri-tourism farms. The study also notes that the age of the farmer has an inverse relationship on gross sales. Islam et. al., (2011) studies show that length of time of operation has significant positive effect on business success of SMEs. Their results again show that SMEs that have operated for longer periods have been more successful in comparison to those who have been in operation for a shorter period.

5.2. Ownership Characteristics of agro-processing firms in Ghana
This section presents some socio- demographic characteristics of firm owners in the study area.

From the results presented in Table 5.1 below, 23.2% of the total firms studied are male-owned. Of these, 44.4% have no formal education whiles 8.1% have primary education. However, only 9.7% have tertiary education. Female firm owners constitute 76.8% of the total firms surveyed. Surprisingly, 89% of these female firm owners have no formal education whereas less than one per cent has attained higher or tertiary education.

The results confirm that most of the food crops agro-processing firms are female-owned with less formal education compared to the male owners. The difference in educational levels could play off in their access to information, credit, training and other resources, as well as in innovation and technology adoption.

Table 5.1: Agro-processing Firm Ownership Characteristics (percentages)

Gender	Educational Level					Overall
	None	Primary	J.H.S	S.H.S	University / higher education	
Male	44.4	8.1	16.1	22.6	9.7	23.2
Female	89.0	7.2	2.4	1.0	0.5	76.8
Total	*78.7*	*7.4*	*5.6*	*5.9*	*2.6*	*100*

Source: Field Survey 2012

The results above may supports a study by McNeil (2005) who estimated that about 90 percent of the activities involved in processing agricultural products in Ghana are done by women and this makes them key stakeholders in improving the agro-processing sector. Okorley (1999) agrees that in Ghana, the role of women in food production, processing and marketing has become more important.

The age of the firm owner is a very important characteristic. The minimum, maximum and mean age of the firm owners are 20, 82 and 46.6 years respectively. Table 5.2 presents the distribution of the various agro processing firms according to age and gender. It can be seen from Table 5.2 below that majority of the firm owners (74%) are 40 years and above. This implies that most of the younger people in Ghana are not involved in food crops agro-processing. This may result from the fact that most Ghanaian youth do not have access to land, finance and the required skills or the sheer interests in this business. The capital involved in setting up an agro processing firm with modern technology is high and the fact that young people are not willing to work with the indigenous and outdated tools that are being used in agro processing firms in Ghana could be some of the reasons for the low involvement of the youth and young adults in agro-processing.

It is also worth noting from Table 5.2 that 79 percent and 72 percent of the males and females constitute the adult group (more than 40 years) respectively. These results from the sample survey suggest that there are relatively more young females in agro-processing than there are males.

Table 5.2: Distribution of Age of Agro-processing firm Owners by Age

Gender	Age of respondents					Overall
	20-29	30-39	40-49	50-59	60+	
Male	2(3)	11(17)	24(38)	15 (24)	11 (17)	63(23.2)
Female	11(5)	48(23)	60(29)	63(30)	27(13)	209 (76.8)
Total	*13(5)*	*59 (22)*	*84 (31)*	*78 (29)*	*38 (14)*	*272 (100)*

Source: Field Survey 2012 *NB: Figures in parenthesis are percentages*

Table 5.3 presents the distribution of the various agro processing firms with respect to their gender and the category of item processed. From the sample survey, whilst 9.5 percent of males venture into alcoholic beverage processing less than 1 percent of the females process alcoholic beverage. On the contrary, whilst 10.5 percent of the females are into cereal production, only 3.2 percent of the males process cereals. This could be explained from the social perceptions and expectations that sometimes influence occupational choices. For example, having a business venture in alcoholic beverages is perceived to be a man-business whilst cereal processing is a business mostly regarded to be for women.

The study shows that majority of males and females firm owners are into processing of oil and oil by product. But in relative terms, there are more female firm owners involved in the processing of oil and oil-related products than the male counterpart. As mentioned in chapter 4, this could be because of the increased demand for oil products.

Table 5.3: Distribution of various agro-processing firms by gender

Sex	Crop Category							Overall
	Roots and tubers	Cereals	Vegetables	Oil and Oil by-products	Soaps and cosmetics	Alcoholic beverages	Non-alcoholic beverages	
Male	10(15.9)	2(3.2)	2(3.2)	41(65.1)	1(1.6)	6(9.5)	1(1.6)	63(23.2)
Female	56(26.8)	22(10.5)	0	113(54.1)	14(6.7)	1(0.5)	3(1.4)	209 (76.8)
Total	66(24.3)	24(8.8)	2(0.7)	154(56.6)	15(5.5)	7(2.6)	4(1.5)	272 (100)

Source: Field Survey 2012

NB: Figures in parenthesis are percentages

The various sampled firms are grouped in 4 decades according to their year of establishment. Their distribution in terms of scale of the firm is presented in Table 5.4.

Table 5.4: Firm's Scale and Year of Establishment in percentages.

Year of Establishment	Firm Category				Overall
	micro	very small	small	Medium	
1st Decade (1st -10th Year)	100	0	0	0	38 (14)
2nd Decade (11th to 20th Year)	187 (86)	14 (6)	9 (4)	8 (4)	218(80.14)
3rd Decade (21st to 30th Year)	6 (46)	3 (23)	4 (31)	0	13 (4.8)
4th Decade (31st to 40th Year)	1 (33.3)	1 (33.3)	1 (33.3)	0	3 (1.1)
Total	232 (85.3)	18 (6.6)	14 (5.1)	8 (2.9)	272 (100)

Source: Field Survey 2012

NB: Figures in parenthesis are percentages

As can be seen from Table 5.4, most of the firms surveyed, about 80.14 percent are in their second decade and only 3 of the firms are in their 4th decade. Of the 80.14 percent 2nd decade firms, 86 percent of them are micro, 6 percent are very small firms and only 8 percent are either small or medium scaled firms. All the firms in their beginning years (1st Decade) are micro-scale firms. Also, 31 percent of the 3rd Decade firms are small scale. Despite the fact that over a period of time, all things being equal, agro-processing firms in Ghana, should grow from micro to become medium scale and subsequently large scale, it is seen from the results in Table 5.4 that most firms are not growing thus most firms are in the first and second years. This could also be an indication of lack of transitional planning and business continuity plans to mentor the younger generation to continue with the operation of the firms. This could also result from the fact that most of the businesses are more than 40 years (Table 5.2) and are using indigenous methods to grow the businesses which collapse in either the first or second decade. Firm managers are neither able to take micro firms to the next decade nor to a higher firm scale.

5.3. Building Structure of Agro processing firms in Ghana

The nature and ownership status of agro-industrial building complexes are described in this section. The building structures for the sampled agro firms are either storey/complex buildings, simple single structures which are either self-owned, rented or family-owned (Table 5.5)

It can be seen from Table 5.5 below that about 94 percent of the structures of the sampled firms are simple single structures, whilst, 6.5 percent had complex or storey buildings. Majority (56.2 percent) of the structures were rented whilst about 40 percent were either owned by the firm or the family. The results also show that whilst about 58 percent of the single structured buildings were rented, 70.6 percent of the storey buildings were owned by the firm.

Table 5.5: Nature of the agro-industrial complex and their respective ownership status

Nature of the agro-industrial complex	Ownership status of agro-industrial complex				Total
	owned	rented	family owned	others	
complex or storey building	12 (70.6)	5 (29.4)	0	0	17 (6.5)
Simple single structure	81 (33.3)	141 (58.1)	11 (4.5)	10 (4.1)	243(93.5)
Overall	93 (35.8)	146 (56.2)	11 (4.2)	10 (3.8)	260 (100)

Source: Field Survey 2012

NB: Figures in parenthesis are percentages

Table 5.6 presents the nature of the agro-industrial complex for the various firm scale. It can be seen that whilst 50 percent of the medium scale firms used a complex or storey building about 94 percent of the micro firms used simple single structures. However, contrary to expectations, about 82 percent of the complex/storey buildings are used by micro-scale firms. From observations during the field studies, some of the micro-scale firms operate from their rented homes in complex/storey buildings. About 2 home based micro-scale firms operated by recent retirees are also located in their complex/storey building homes.

Figure 5.1 Some Simple Agro Processing Structures

Source: Field Survey 2012

Table 5.6 Nature of agro-industrial complex by firm scale

Firm Scale	Nature of the agro-industrial complex		Total
	complex or storey building	Simple single structure	
micro	14 (6.2)	212 (93.8)	226
very small	0	17 (100.0)	17
small	0	11 (100)	11
medium	3 (50.0)	3 (50.0)	6
Overall	17 (6.5)	243(93.5)	260

Source: Field Survey 2012

NB: Figures in parenthesis are percentages

Table 5.7 shows that only the medium scaled firms interviewed do not rent or use family structures for their operation. They all operate from building structures owned by the firm. This may indicate that medium scale firms have the financial capacity to invest in buildings as part of the corporate assets.

Table 5.7 Ownership status of agro-industrial complex by firm scale

Firm Scale	Ownership status of agro-industrial complex				Total
	owned	rented	family owned	others	
micro	77 (34.1)	130 (57.5)	9 (4.0)	10 (4.4)	226
very small	5 (29.4)	10 (58.8)	2 (11.8)	0	17
small	5 (45.5)	6 (54.5)	0	0	11
medium	6 (100)	0	0	0	6
Overall	93 (35.8)	146 (56.2)	11 (4.2)	10 (3.8)	260

Source: Field Survey 2012

NB: Figures in parenthesis are percentages

5.4. Management capacity and Skill of Agro processing firms in Ghana

Most agro processing firms in Ghana do not have the basic skills in operations management, resource mobilization, and financial management. The study notes this as one of the critical constraints facing agro-processing firms. As shown in Figure 5.2, apart from 21 percent of the owners who perceive they do not lack any skill, the remaining (79 percent) of firm owners claim they do not have adequate capacity in several areas. About 35 percent of the firm owners claim they lack the skills to mobilize resources and to manage human resources. This is very alarming especially in a country like Ghana where resources are abundant and labor is not expensive. This substantiates the findings by Mugera (2012) that agribusiness managers face the challenge of managing their employees in an effective and efficient manner.

Whilst 19 percent of the firms interviewed indicated they do not have technical skills or are not well trained to handle advanced technology for their operations, 15 percent of the firms consider their skill constraints to be in the area of financial management. The lack of financial management skill is evident in the fact that from the data, 57 percent of the firm owners do not keep good financial records.

Figure 5.2: Skill Constraint of the various firms

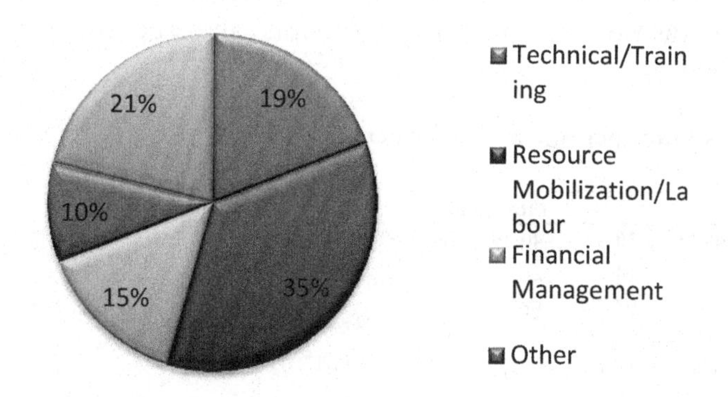

Source: Field Survey 2012

The skill constraints of the various product categories are analyzed and presented in Table 5.8 below. Out of the firms who claimed they lack skills in resource/labor mobilization, 62.5 percent are oil and oil by-product processors, 29.2 percent are either root and tuber processors or cereal processors, 6 percent are from soaps and cosmetics and 2 percent are either vegetable or non-alcoholic beverage processing. The results suggest that a lot more resource mobilization skills is needed for oil and oil products processing firms.

Table 5.8: Firm owners Skill Constraints by major products of the firms

	Crop Category							Over-all
	Roots and tubers	Cereals	Vegetables	Oil and Oil by - products	Soaps and cosmetics	Alcoholic beverages	Non-alcoholic beverages	
Technical/Training	18(34.6)	5(9.6)	1(1.9)	19(36.5)	4(7.7)	3(5.8)	2(3.8)	52(19.1)
Resource Mobilization/Labor	14(14.6)	14(14.6)	1(1.0)	60(62.5)	6(6.3)	0	1(1.0)	96(35.3)
Financial Management	15(37.5)	3(7.5)	0	17(42.5)	4(10.0)	0	1(2.5)	40(14.7)
Other	8(30.8)	0	0	16(61.5)	0	2(7.7)	0	26(9.6)
No Skill Constraint	11(19.0)	2(3.4)	0	42(72.4)	1(1.7)	2(3.4)	0	58(21.3)
Total	66(24.3)	24(8.8)	2(0.7)	154(56.6)	15(5.5)	7(2.6)	4(1.5)	272(100)

Source: Field Survey 2012

NB: Figures in parenthesis are percentages

Interestingly, the study shows that both younger firm owners and older firm owners have similar skill constraints. This is presented in Table 5.9 which shows the distribution of skill constraints and age of agro-processing firm owners.

Table 5.9: Skill Constraints per Age of firm owners

Skills Contraints	Age of Respondents					Overall
	20-29	30-39	40-49	50-59	60+	
Technical /Training	1 (1.9)	14 (26.9)	24 (46.2)	6 (11.5)	7 (13.5)	52(19.1)
Resource Mobilization/Labor	7(7.3)	17(17.7)	23 (24.0)	32(33.3)	17(17.7)	96(35.3)
Financial Management	0	7(17.5)	12(30)	16(40)	5(12.5)	40(14.7)
Other	5(19.2)	10(38.5)	8(30.8)	1(3.8)	2()	26(9.6)
No Skill Constraint	0	11(19.0)	17(29.3)	23(39.7)	7(12.1)	58(21.3)
Total	13(4.8)	59(21.7)	84(30.9)	78(28.7)	38(14.0)	272(100)

Source: Field Survey 2012
NB: Figures in parenthesis are percentages

Surprisingly, the sample survey suggests that, whilst 51 percent (about half) of the agro-processing firm owners who lack adequate skills in resource mobilization are 50 years and older 48.3 (about half) percent of the respondent who indicated they did not have any skill constraint are 50 years and below. This implies that solutions to the constraints that older agro-processing firm owners face are not sought for by the younger generation.

5.5. Employments in Agro-Processing Firms in Ghana

According to Mugera (2012), employees are strategic resources for agribusiness firms to achieve sustained competitive advantage. Human resources are one of the crucial strategic assets in agribusiness. An important feature of agro-processing industries is that they are a major source of employment and income- especially in rural communities. Agro processing can contribute to the relief of rural unemployment and underemployment. According to a report by the Food and Agriculture Organization of the United Nations (FAO), agro-processing industries (food, beverages and tobacco) typically employ about 10 percent of the total labor force in manufacturing in the developed countries and around 20 to 30 percent in the developing countries. Also, the highest shares of employment in agro-processing firms are found in Africa (FAO, 1997).

As noted in chapter 4, Ghana's agro processing firms are dominated by micro firms that employs from 1 to 5 persons, most of the firms are involved in oil and oil by product processing, and the Western Region has more agro processing firms. The agro processing firms in Ghana also contribute to the economy by its employment capacities. The 272 firms surveyed in this study employ a total of 1,520 persons. The average employment per firm is about 6 people ranging from 1 person to 100 people. Figure 5.3 presents the employment of agro processing firms in the various regions in Ghana. From Figure 5.3, the Brong Ahafo region, Western region and Northern region employs 27 percent, 23 percent and 18 percent of the total labor force recorded from the survey data respectively. With the assumption that more labor increases productivity all other things being equal, it could be implied that the agro-processing firms in these three regions (Brong Ahafo, Western and Northern) could improve their productivity since they employ more labor.

Figure 5.3: Total number of People Employed in the various regions

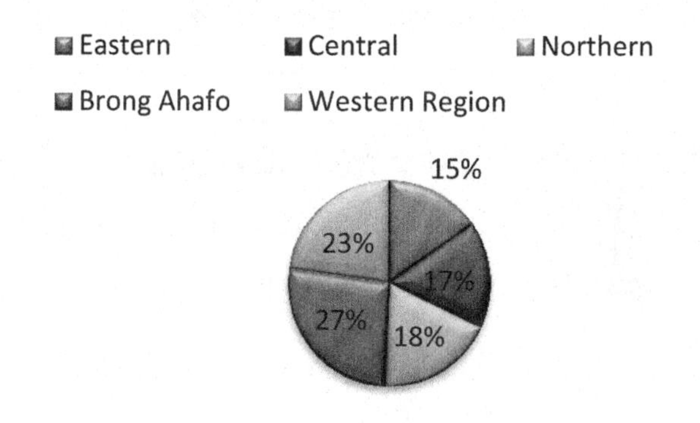

Source: Field Survey 2012

It is also obvious that the dominant agro processing firms would employ more people. This is illustrated in Figure 5.4 below.

Figure 5.4: Total Number of People Employed in the various Products Categories

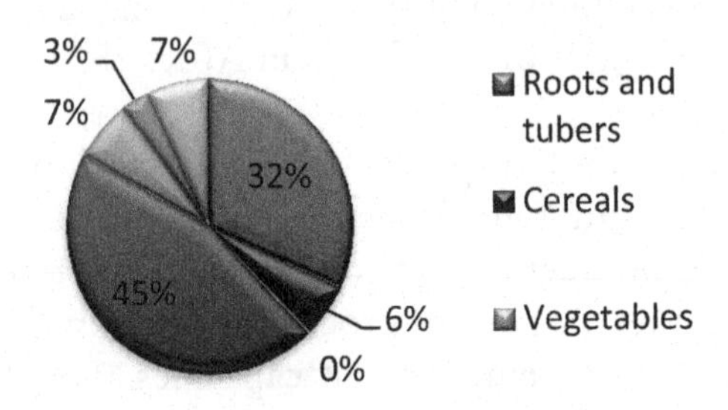

Source: Field Survey 2012

From Figure 5.4, 52 percent of the total agro processing employees are in the oil and oil-by product processing. This could also be because the processing of oil is quite labor intensive and involves several activities as illustrated in Chapter 3. In Figure 5.5 below, we see that the micro agro-processing firms employ up to 48 percent of the total agro-processing employees. Obviously, majority of the firms interviewed are micro-scale firms and this indicates the importance of these firms for employment, household income and inclusive growth in the nation.

Figure 5.5: Total number of People Employed by Firm Category

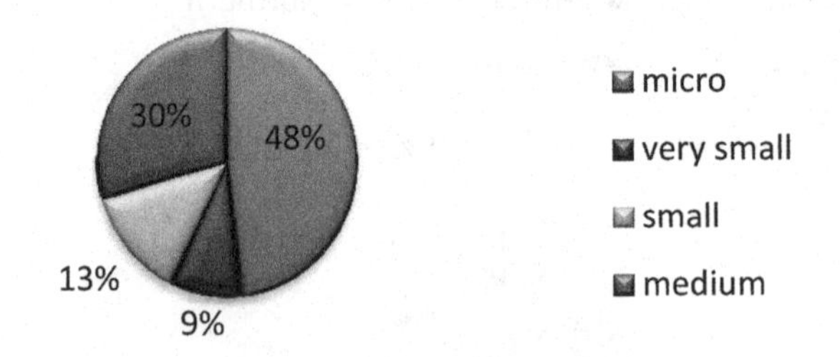

Source: Field Survey 2012

Though the small and very small firms have higher distributions percentages of 5 percent and 7 percent respectively (Figure 4.1), compared to the medium of 3 percent, the medium scale agro processing firms employ 30 percent of the total agro processing firms. Table 5.10 below presents the average number of people employed in each of the firm scale. On the average whilst micro firms employ about 3 people, the medium firm scale employs about 56. This result attests to the fact that as firm increases in scale they tend to employ more labour.

Table 5.10: Average number of people employed by different firm scales

Firm Scale	Average number of people employed
micro	3.16
very small	7.44
small	14.57
medium	56.00

Source: Field Survey 2012

5.6. Conclusion

This chapter discussed the major ownership and employment characteristics of food crops agro-processing firms in Ghana, to provide information that will help address the critical knowledge gap in these issues. It has presented the characteristics of agro processing firm owners, their skill constraints and their employment capacities. The results show that females constitute the greatest proportion of food crop agro-processing owners. Since, most of the firms in the food processing industries are micro to small businesses, it is a general indication that majority of these women operate micro to small enterprises.

Another characteristic of the owners of most food crop agro-processing firms is their low level of formal education. This may affect their training and skills acquisition as well as information search and adoption of technology. It is evident that over 80 percent of the firm owners interviewed lack adequate capacity in areas such as resources mobilization and financial management. This finding should underlie the need for periodic training and other capacity development activities for agro-processing firms in order to improve their operational and managerial efficiency.

Since women are the dominant stakeholders in Ghana's agro-processing sector, for the sector to grow and contribute significantly to socio-economic development, there will be the need for capacity enhancement activities that target female firm owners. Extension agents could organize systematic management training for agro-processing owners especially in resource mobilization, financial management and modern manufacturing techniques in order to realize optimum contribution of the sector to the Ghanaian economy. The latest technology developed by the research institutions should be effectively communicated to fill up gap through organizing seminars and conferences of the potential entrepreneurs. Firm owners could be exposed to recently developed technologies through organizing exhibition and mini technological fair.

The agro-processing firms in Brong Ahafo, Western and Northern Regions employ the majority of the labor force in the agro-processing industry. Micro- and medium-scale firms have relatively larger employment capacity. They employ majority of the labor force in this sector. The predominance of micro-scale firms and their employment potentials show their significance for improved household income, inclusive growth and diversification of rural economies in Ghana. However, majority of the micro-scale firms are only in the first decade of establishment which gives an indication of low survival rate for these firms and the unfortunate lack of transitional planning and business continuity to improve their scalability and sustainability. Firm owners may need some level of training in succession planning and business continuity strategies.

CHAPTER SIX: FOOD CROPS AGRO-PROCESSING OPERATIONS

6.1 Introduction

Globally, food crop agro-processing operations are gradually evolving. Growing complexities in technologies and increasingly sophisticated consumers with changing preferences regarding nutrition, health and safety demands are driving improvements in food crop agro-processing operations. Also, growing population, changing work roles of women with less time for domestic cooking and the growing middle income class has contributed to the shift in consumers' preference from unprocessed food products to processed ones.

6.2 Installed and Operating Capacities

It will be appropriate to understand the capacity of agro-processing firms in Ghana to respond to these changes in market trends and their ability to take advantage of the increasing demand for processed food products. This section will discuss the installed capacity as well as the operating capacity for small and medium-scale food crops agro-processing in the nation.

A firm's installed capacity - defined in terms of the maximum level of output that it could produce in a given time period may influence the short term decision of the firm. Manufacturers of food crop agro-processing equipment often specify the installed capacity. However, this could be lower depending on the location, installations, electrical connectivity, and operator's inventory of resources including labor, skills and knowledge. Processing equipment may also have their minimum as well as maximum capacity at which an appropriate installation could be operated for a sustained period without causing damage to it. Therefore, the selection and installation of the right capacity is an important decision to be taken by the agro-processing firm. Ideally, the installed capacity of food crop agro-processing facility should be such that it is able to meet customers demand during periods of peak demands as well as off-seasons.

During the field survey, most of the food crop agro-processing firms, particularly the micro, very-small and small food crop agro-processing firms, could not specify their installed capacity mainly because they had no installed facility or equipment but resort to traditional agro-processing methods. Another challenge was the use of crude indigenous methods to measure capacity. Thus, the lack of standardization in capacity measurement across firms involved in processing of the same product type also posed serious setbacks to capacity measurement. This notwithstanding, some efforts were made to obtain the monthly installed capacity of some firms. These have generally been grouped according to the type of commodity the firm produces for easy understanding.

As shown in Table 6.1, firms involved in the processing of vegetable products generally have high average installed capacity in terms of kilograms of commodity produced per month. This is probably due to the high capacity of the few firms specializing in vegetable products. Firms in the soaps and cosmetic sector have low capacity relative to firms in the other product categories. In terms of measurement in gallons, non-alcoholic beverage producers such as producers of fresh fruit juices have relatively high installed capacity, producing on average 700 gallons per month. Oil producing firms come next with average installed capacity of 260.4 gallons per month. However, the minimum installed capacity of non-alcoholic beverage and oil producing firms suggests that these are installed capacities of medium- sized firms in the sector.

Table 6.1: Installed Capacity of some Agro-processing Firms in Ghana

Commodity produced by firm	Min	Max	Mean
Vegetables (Kg)	1650.0	2400.0	4050.0 (81.0)
Roots and tubers (Kg)	750.0	1400.0	1200.0 (24.0)
Cereals (Kg)	400.0	1000.0	700.0(14.0)
Soaps and cosmetics (Kg)	80.0	400.0	194.3 (3.9)
Oil products (gallons)	200.0	320.0	260.4
Alcoholic beverages (gallons)	32.0	60.0	45.0
Non-alcoholic beverages (gallons)	630.0	720.0	700.0

NB: Figures in parenthesis are conversions into number of 50Kg bags or sacs.
Source: Field Survey 2012

The operating capacity, explained as the percentage of the firm's total possible production capacity currently being utilized, determines the cost of production as it has influence on fixed costs per unit of product produced and the size of firms. The level at which a firm operates is in turn determined by the extent of competition and market share, demand and preferences, sales trend and the availability of raw materials. In Ghana, availability of raw materials to feed these food crop agro-processing facilities often varies according to food crop production cycles. Therefore, installed agro-processing facilities may remain underutilised during off-seasons and sometimes over utilised during peak demands.

The operating capacity also has influence on profitability and the ability to respond to consumers demand in the short run production period. Therefore, under-utilisation as well as over-utilisation or full capacity utilisation has their merits and demerits. Under-utilisation of capacity is unlikely to be desirable in the short term as the higher fixed cost per unit and eventual higher unit costs of final product will make it difficult to compete. However, there may be less stress for employees than if they were working at full capacity. A firm may also have more time for maintenance and repairs and for staff training, to prepare for an upturn in trade or to cope with new orders.

On the other hand, full capacity utilisation or over-utilisation of a firm's capacity may reduce fixed costs per unit of production, and consequently total cost per unit. However, there may not be enough time for routine maintenance, so machine breakdowns may occur more frequently and orders may be delayed. It may also present a challenge to meet new or unexpected orders so the business cannot grow without expanding its scale of production and employing additional resources. For example, if the firm is operating over its capacity, staff may be stressed leading to increased mistakes, absenteeism and high labor turnover. If the factory space is overcrowded, work may become less efficient due to the untidy working conditions and it may be necessary to spend more on staff overtime to satisfy orders.

The operating capacity of food crop agro-processing firms may be empirically estimated as:

$\frac{\omega_t}{\aleph_t} \times 100\%;$ (6.1)

Where:

ω_t = actual output per time period t (t = monthly) and;

$\aleph_t$= maximum possible output per time period t (t=monthly)

Firms operating capacity are desirable between 80 to 90%, as fixed costs per unit will be relatively low. With this operating capacity, there is also some allowance to meet new orders or to carry out maintenance and training. This is referred to as the capacity cushion (Greasley, 2007). Figure 6.1 illustrates the operating capacity of food crop agro-processing firms in Ghana. From the figure, 22.1 per cent of the food crop agro-processing firms surveyed operate below 50 per cent of their installed capacity. This may be a source of increasing cost of per unit production of many locally produced food products. Firms in this category are likely to fold up due to their inability to meet increasing variable cost of production and competition. Low demand, small market share and their inability to price their products at competitive levels may be another reason. About 62%) of the firms are operating in the range of 61 to 70 per cent of their installed capacity. Firms in this category may be able to meet variable cost but cannot cover fixed cost in the short-run production period; hence, may remain in operation for some time. These firms may also benefit from short-run increases in demand.

Figure 6.1 Operating Capacity of Food Crop Agro-processing Firms in Ghana

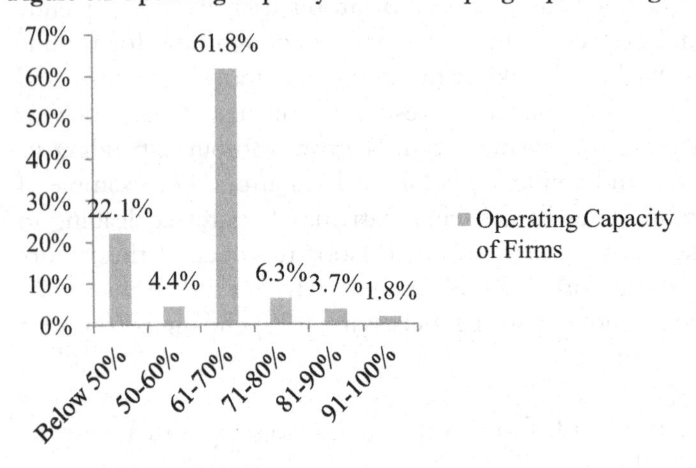

Source: Field Survey 2012

Another category is those in the range of 71-80%. These groups of food crop agro-processing firms form the third highest in the industry. These firms though are underutilizing their capacity, were operating close to the desired operating capacity. These firms therefore may be meeting all their variable cost as well as their fixed cost. However, they require minimal support to operate efficiently. Only 3.7% of the food crop agro-processing firms interviewed are operating in the ideal utilization capacity range of 81-90%. Also, 1.8% of the firms are operating at full capacity. These firms may not be able to respond to short-run increases in consumer demands. Increases in wear and tear due to lack of regular maintenance may also be associated with these firms.

6.3 Level of Technology Use

Efficient food crop agro-processing operations rest on effective technological and innovative systems. Increasingly, the quality, durability and branding of processed food products are being shaped by improved processing and packaging technologies. Therefore, the extent of technology use is an important factor influencing the quantity and quality of processed food product. The kind of technology used can also reduce cost by saving input use such as energy and power and reduce drudgery.

These technologies can broadly be classified as manual tools or machines, and motorised equipment. Most firms combine both manual and motorised equipment but at varying proportions. It was observed during the field survey that most of the micro and very small-sized food crops agro-processing firms rely solely on manual tools or on a higher proportion of manual tools and to a limited extent on motorised machines Conversely, medium and large scale agro-processing firms depend largely on motorised equipment but at varying proportions.

Figure 6.1 A picture of some manual tools being used for agro-processing

A milling machine — Squeezing palm oil from the boiled palm nut.

Source: Field Survey, 2012

Table 6.2: Equipment used by Food Crop Agro-processing Firms in Ghana

Micro	Very small	Small	Medium
Cooking pot	Cooking pot	Cooking Pot	Factory packing line
Pans	Head pan	Pans	Drying oven
Head pans	Bucket	Frying machines	Storage chambers
Drum/ barrels	Bowls		Cutting machines
Plastic containers	Drums		Drying equipment
Bowls			Gas oven
Frying Spoons			Stand by generator
Basin			Water tankers
			Storage tanks
			Pick up vehicles
			Official cars

Source: Field Survey 2012

As shown in Table 6.2, most of the micro, very-small and small food crop agro-processing firms had low level of technology use as they resort to traditional methods of processing using primary or manual tools such as cooking pots and pans. Medium-sized food crop agro-processing firms; however, use reasonable level of advanced technology.

6.4 Efficiency of Conversion

Food crop agro-processing firms have several competing needs. New firms get started each day and thousands of new products are introduced each year. Consumers now demand processed food products that are convenient and of highest quality in terms of flavor, color and texture. This presupposes that for a firm to remain competitive and maintain its position as a key industry player, then it must be efficient in the conversion of its inputs into quality products. Rising consumerism therefore calls for increasing efficiency in the processing of food crops by food crop agro-processing firms if they are to meet these changing consumer needs. Thus, in a competitive market, efficiency is important to firm's existence. Therefore, it is often appropriate to measure the efficiency of firms in an industry to inform policy.

The efficiency of the transformation processes for firms in an industry can generally be estimated by the units of output per unit of input use, from the economic perspective. Mathematically, this is specified as:

$$\frac{Z_t}{I_t} \times 100\% \qquad (6.2)$$

Where:
Z_t = Total output of the food crop agro-processing firm over a time period t, (t=weekly, biweekly, monthly or annually);
I_t = Total inputs used by the food crop agro-processing firm over a time period t (t=weekly, biweekly, monthly or annually)

Output and input from a firm could be valued in monetary terms or measured in terms of volumes or kilograms of output produced from a unit of input use. Food crop agro-processing firms operating below 100% are considered to be less efficient. From Table 6.3, the mean efficiency of all food crop agro-processing firms considered in the study is 28.2%. This reflects low efficiency of the food crop agro-processing firms in Ghana and the possible source of increasingly higher price per unit of production. About 52% of the firms interviewed have conversion efficiency of less than 25% whereas 44.9% achieve 25 to 50% conversion efficiency. Only 3.3% operate with 50 to 75% conversion efficiency. This implies that most of the food crop agro-processing firms are less efficient in their input use.

Table 6.3 Distribution of Conversion Efficiency of Food Crop Agro-processing Firms in Ghana.

Efficiency levels (%)	Frequency	Percent	Mean efficiency (%)
< 25	141	51.8	
25-50	122	44.9	28.2
50-75	9	3.3	
Total	272	100.0	

Source: Field Survey 2012

The source of this inefficiency in the conversion of input into output can be considered from the type of agro-processing firm. From Table 6.4, conversion efficiency increased with firm size. This proposes that conversion inefficiency may be due to diseconomies of scale and scope. From the result, 56% of micro-sized food crop agro-processing firms have conversion efficiency of less than 25% whereas 42.2% and 1.7% have efficiency levels between 25 to 50% and 51 to75% respectively.

Similarly, for very small food crop agro-processing firms, 38.9% have conversion efficiency of less than 25% whilst 61.1% operates within 25 and 50% efficiency of conversion. The result indicates that majority of small and medium-sized food crop agro-processing firms in Ghana have conversion efficiency in the range of 25 to 50%. However, unlike the micro, very small or small food crop agro-processing firms, none of the medium-sized food crop agro-processing firms operate below 25% conversion efficiency. This may be due to the relative diversification of products where 'waste' from one output is used as an input for another product. This maximizes the output from unit input or raw material.

Table 6.4 Conversion Efficiency by Types of Food Crop Agro-processing Firms in Ghana.

Firm type	Efficiency levels			Mean Efficiency
	Below 25%	25-50%	51-75%	
micro	130(56.0)	98(42.2)	4(1.7)	26.7
very small	7(38.9)	11(61.1)	0(0.0)	30.4
small	4(28.6)	8(57.1)	2(14.3)	37.2
medium	0(0.0)	5(62.5)	3(37.5)	51.8

Source: Field Survey 2012

In terms of product-specific food crop agro-processing firms, those producing alcoholic beverages have the greatest mean conversion efficiency. Food crop agro-processing firms involved in the processing of vegetables and non-alcoholic beverages comes second and third respectively in terms of mean conversion efficiencies. Cereals as well as oil producing food crop agro-processing firms were two of the food crop agro-processing firms with the least conversion efficiency. This observed trend suggests that the type of raw material used by the firm has influence on the firm's conversion efficiency.

Thus, besides demand and market conditions, the proportion of the raw material converted into finished products by the agro-processing technology is a major determinant of efficiency of conversion. For instance, besides the existence of large market for alcoholic beverages, there is also minimal waste associated with its production relative to other products; hence giving alcohol producing firm's high conversion efficiency. In terms of percentage distribution, whereas majority of the firms in the alcohol producing enterprise operate in the range of 25 to 50% as also found with all of the firms in the vegetable industry, majority of the firms in the roots and tubers, and cereal subsector operates below 25% conversion efficiency.

Table 6.5 Product-specific Conversion Efficiencies

Product type	Efficiency levels			Mean Efficiency
	Below 25%	25-50%	51-75%	
Roots and tubers	36(54.5)	29(43.9)	1(1.5)	28.36
Cereals	12(50.0)	12(50.0)	0(0.0)	26.53
Vegetables	0(0.0)	2(100.0)	0(0.0)	34.32
Oil and Oil by-products	83(53.9)	64(41.6)	7(4.5)	27.76
Soaps and cosmetics	8(53.3)	7(46.7)	0(0.0)	28.11
Alcoholic beverages	1(14.3)	5(71.4)	1(14.3)	36.9
Non-alcoholic beverages	1(25.0)	3(75.0)	0(0.0)	33.95

Source: Field Survey 2012

Regional trends suggest that the location of a food crop agro-processing firm determines the efficiency of conversion. As depicted in Table 6.6, food crop agro-processing firms located in the Eastern region operates with relatively high conversion efficiency whereas firms located in the Western region have relatively low efficiency of conversion.

Table 6.6 Regional Distribution of Conversion Efficiency of Food Crop Agro-Processing Firms in Ghana

Region	N	Minimum	Maximum	Mean
Eastern	12	19.93	67.57	33.26
Central	63	13.83	53.00	30.13
Northern	41	15.67	65.03	27.66
Brong Ahafo	60	13.13	52.14	27.84
Western	96	14.15	52.37	26.73

Source: Field Survey 2012

6.5 Products and Services

Modern food crop agro-processing operations strive to provide products or services that meet consumer expectations. This may involve biological, physical and chemical reactions or processes that prevent the growth of microbes, change the physical structure or alter the chemical composition of the food product. Methods such as refrigeration, freezing, drying, pasteurization, sterilization (canning), grinding, heating, mixing, membrane processing and fermentation are often used to achieve this purpose. However, most food crop agro-processing operations may comprise a series of these physical, biological and chemical processes that can be broken down into a number of basic operations often termed as unit operations. Unit operations may cut across activities such as:

Fluid flow- This involves moving a fluid product from one point to another with varying degrees of turbulence);
Heat transfer - Heat is either removed or added and includes heating, cooling, refrigeration or freezing); contact equilibrium processes or mass transfer (which involves transfer of molecular species from or to a product such as distillation, gas absorption, crystallization, membrane processes, drying, evaporation); mechanical separation (such as filtration, centrifugation, sedimentation, sieving); size adjustment (either through size reductions methods such as slicing, dicing, cutting, grinding or size increase through aggregation, agglomeration, gelation); mixing (to make homogenous blends such as solubilizing solids, preparing emulsions or foams or dry blending of ingredients) and fermentation (ref). Firms may use a single or multiple unit operations in the processing of their products. Thus, whiles some unit operations can stand alone others may depend upon coherent physical, chemical and biological principles.

Food crop agro-processing firms involved in the provision of primary products basically provide services to other food crop agro-processing firms or individuals by performing one or more of the unit operations. Primary products require further processing operations to improve their appearance, taste and sometimes the nutritional value. Secondary and convenient food crop agro-processing products are prepared food stuffs that can be sold as hot, ready-to-eat dishes; at room-temperature, shelf-stable products; or as refrigerated or frozen products that require minimal preparation (Anderson and Deskins, 1995). Growing affluence of today's consumers therefore implies that food products that are convenient to use are carefully engineered through a combination of these unit operations.

As shown in Table 6.7, 98.2% of the food crop agro-processing firms in Ghana produce primary commodities such as corn or cassava dough, milled or polished rice and other products. Among the primary products, palm oil and gari dominates. Recent novel and convenient products such as 'Garimix' and 'Neat fufu' that are making inroads into the market are likely to significantly reduce the market share of those food crop agro-processing firms involved in processing of primary products such as corn dough. Secondary products only form 1.8% of products by food crop agro-processing firms but most (80%) of these products are fruit juice.

Table 6.7 Products Produced by Agro-processing Firms in Ghana

Product classification	Overall	Products	Frequency	Percent
		Cassava dough	6	2.2
		Corn dough	1	0.4
		Processed okro	1	0.4
Primary products	267 (98.2)	Palm kernel oil	10	3.7
		Cassava powder	1	0.4
		Milled rice	23	8.6
		Shea butter	14	5.2
		Gari	59	22.1
		Palm oil	140	52.4
		Local gin / Akpeteshie	7	2.6
		Groundnut powder	4	1.5
		Chilled cabbage	1	0.4
		Total	267	100
Secondary products	5 (1.8)	Fresh fruit juice	4	80
		Soap	1	20
Total	272 (100.0)	Total	5	100

Source: Field Survey 2012

6.6 Internal and External Linkages of Operation

Food crop agro-processing firms may cover only part of their value chain and depend critically on other firms or individuals to complete the production process (Pfeffer and Salancik, 1978). Since food crop agro-processing firms may have inadequate resources, it is common place for part of their operations or value chain to be outsourced to other players, and transact with these actors in the value chain. Internal linkages arise when firms build a network with other actors in the value chain or other food crop agro-processing firms locally located in Ghana or are part of an internal network which stretches beyond Ghana but the firms are of Ghanaian origin. Thus, agro-processing firms may therefore be part of a broader group of multinational companies or may be a single independent or multiple firms all located in Ghana. Internal linkages are developed by the food crop agro-processing firm with objectives ranging from location of the whole or parts of the production process so as to benefit from cheap inputs and taxes; to the development of markets, and the search for assets or resources (IFPRI, 2005).

External linkages are cooperation agreements with the aim of developing and sharing innovations. External linkages can stem from one of two sources: whether food crop agro-processing firms arrange with a multinational institution in Ghana or cooperated with another foreign institution outside Ghana. External linkages occur so that food crop agro-processing firms can exploit potential economies of scale and locational advantages and to realise potential scope advantages by applying innovations and skills generated in one sub-unit of their firm elsewhere in the organization.

Internal and external linkages therefore provide opportunities for intelligence and external control (Burt, 1992). They are also crucial to the discovery of opportunities (Amabile, 1992), building partnerships, expansion and gathering of resources for the formation of the new organization (Aldrich and Zimmer, 1986). Although, the challenge of strengthening, diversifying, and exploiting cross-border linkages within food crop agro-processing industry has emerged as one of the dominant themes in the field of international business today, Ghanaian food crop agro-processing firms are yet to take advantage of these external linkages. Firms interviewed during the field survey had little to report in this section of the questionnaire. From observations, it was realized that many food crop agro-processing firms are unable to develop partnerships or linkages with other cross-border entities partly because of poor management, financial and other book-keeping skills. However, low levels of internal linkages are observed among firms in the food crop agro-processing industry particularly with input suppliers and financial institutions, and support services providers.

Linkages with Input Suppliers and Financial Institutions

Food crop agro-processing firms in Ghana maintain links with their suppliers of raw materials as well as equipment to keep up with new types of equipment coming onto the market and to evaluate their usefulness. These links are also maintained to ensure that after-sales service and maintenance are provided. This occurs sometimes in the form of credit financing to food crop agro-processing firms. From the survey, 37% of the firms have internal linkages with suppliers of inputs such as raw materials and equipment.

Figure 6.2: Distribution of Food Crop Agro-Processing Firms with Internal Linkage in the Purchasing of Inputs

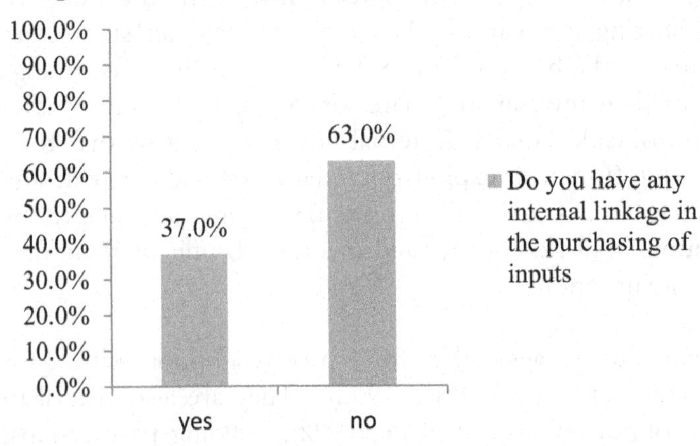

Source: Field Survey 2012

Other firms have links with financial institutions for normal banking transactions such as loan acquisition and issuance of securities.

Relationship with Service Providers

As new technologies get developed and introduced, food crop agro-processing firms would be expected to rely heavily on providers of technical services, unless the firms have their own technical personnel. Also, some of these food crop agro-processing firms rely on rental services for some agro-processing equipment. With the exception of 34.1% who own their processing equipment, the rest rely on service providers for renting on a month-to-month arrangement or for a lease contract which is usually more than six months.

Figure 6.3 Linkages with Equipment Service Providers

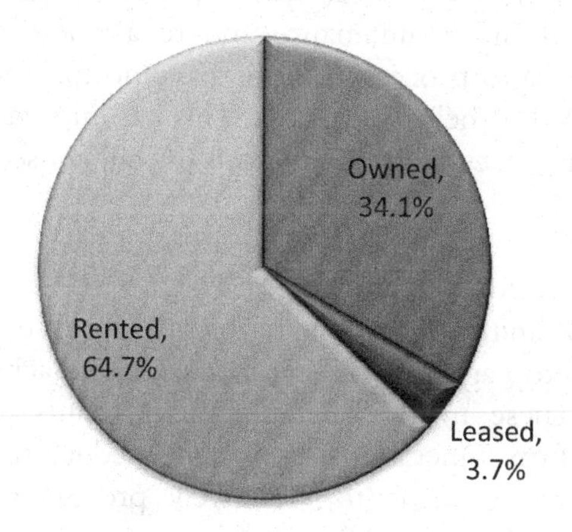

Source: Field Survey 2012

6.7 Operations Management -Repairs and Maintenance of Fixed Assets

A major operational function of any food crop agro-processing firm is to ensure continuous processing of sufficient and quality products at all times. This can only be achieved through regular maintenance and repairs of assets used in the food crop agro-processing operations. Maintenance and repairs of fixed assets help to keep the assets in good working condition or restore it to its previous state of performance. Thus, the reliability of the food crop agro-processing operation depends on regular maintenance of fixed assets of the firm.

Food crop agro-processing firms often lose customers, income, trust and confidence when they continually experience equipment breakdowns. In some instances, injuries and deaths are observed when assets are not maintained or ill-maintained. Therefore equipment used in the processing operation must be routinely inspected and periodically maintained. Ordinarily, maintenance and repairs of fixed assets are activities that the operations manager or the food crop agro-processing firm owners ought to undertake periodically if they are to utilize their fixed assets over their expected service lives. This may be performed by them if they have the expertise themselves or rely on others. However, often times several food crop agro-processing firm owners and managers abrogate this role. Especially in Ghana and other developing economies, firm owners and managers want to see their equipment malfunctioning or broken down before any maintenance and repairs are undertaken. Most of the food crop agro-processing firms have no maintenance and repairs plan or schedule for their fixed assets. This has implication on the quantity and quality of processed products with overall consequences on profitability.

6.8 Operations Constraints

There are a number of limiting factors that affect the quantity and nature of output that a food crop agro-processing firm is able to achieve within a given time period. These are the operational constraints of the food crop agro-processing firm. Operational constraints reduce the ability of food crop agro-processing firms to effectively process and market processed food products. Operational constraints emanate from internal (firm level) or from external sources (macro and global level). At the firm level, some internal operating constraints may include poor maintenance of equipment, poor human resources management including absence of incentive systems to boost employee morale and productivity. It may also include lack of skills and training of the labor force, the existing installed and operational capacity of the food crop agro-processing firm and time constraints. Other firm level constraints include limited access to credit (Chakwera, 1996); lack of appropriate technologies (McPherson, 1996; Mugova, 1996); lack of technological capability; the unreliable supply of raw materials (Mosha, 1983); lack of management skills (Odunfa, 1995); poor product quality control (Jaffee, 1993); and poor markets.

External operational constraints are related to the extent of the market and competition; lengthy procedures; corruption and red tape; legal and illegal fines and harassment. It also includes lack of compliance with agreement by their suppliers, clients and workers; and macro-economic environment, policy, regulatory and institutional frameworks that are not conducive to firm growth and profitability. Trade liberalisation (especially dumping of cheap processed products) and globalization are also posing critical challenges to small and medium-scale food crop agro-processing firms. Although, operational constraints may not affect all firms equally, they are more hostile to micro, very small and small sized food crop agro-processing firms. Many firms have collapsed due to their lack of capacity to handle particularly some of these external operation constraints. The following sections discuss some of these operational constraints and their impacts on food crops agro-processing firms in Ghana.

6.8. 1. Inadequate Human Resource Management skills

Human beings are different from any other resource and hence require special expertise to manage them. Unfortunately, many small food crop agro-processing firms in Ghana are managed by their firm owners who may lack the requisite human resource management skills. As a result, many food crop agro-processing firms are experiencing high labor turnover. Some of these firms also lack effective incentive and motivating systems that can attract and retain talent. Therefore, they are not able to find skilled employees, especially if they are competing with larger companies that can offer more job security and better compensation packages. Most firms also fail to invest in skills training and acquisition. Very few small and medium-scale agro-processors have received formal training in food processing techniques.

6.8.2 Financial Constraints

Food crop agro-processing firms are also saddled with the constraint of access to credit. Limited access to credit affects their capacity to take advantage of rising demand and to adopt and operate modern and high level processing technology. As shown in Figure 6.4, only 37.5% of the food crop agro-processing firms have access to credit. Beyond inadequate access to credit, rising interest rates, short moratorium and short repayment schedules prevent many food crop agro-processing firms from accessing credit to improve their processing operations.

Figure 6.4: Distribution of Firms in terms of their Access to Credit

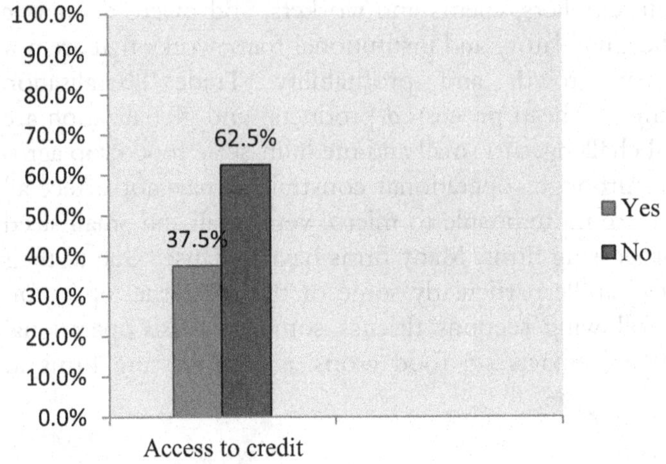

	Yes	No
Access to credit	37.5%	62.5%

Source: Field Survey 2012

6.8.3 Unstable and High Cost of Energy

Erratic power supply and increased cost of energy serves as a major operating constraint to food crop agro-processing firms in Ghana. This is especially so as most firms do not have stand-by power sources such as generators. Frequent power cuts as well as uninformed power outages disrupt processing schedules and the ability of the firm to cope with demand. In some cases, high power surges associated with abrupt and intermittent power outages leads to equipment breakdowns. Food crop agro-processing firms may at times also lose perishable food products. Another worrying trend is the rising cost of energy in Ghana. Lack of access to reliable energy is also one of the major reasons why most of the firms are operating under capacity.

6.8.4 Unreliable Supply

Another important external constraint facing food crop agro-processing firms is supply chain problems. As shown in Figure 6.5, high cost of raw materials and lack of access to raw materials are major hindrances to food crop agro-processing operations in Ghana. Thus, poor access to roads place impediments to accessing raw materials. Supply chain problems is not limited to suppliers of raw materials only but also other manufacturers, distributors, retailers and logistics providers that allow food crop agro-processing firms to get their products to consumers.

Delays or failure by any value chain actor therefore affects the entire operational process of the firm. However, the most important factor is the transportation cost. About 66% of the firm owners interviewed complained of high cost of transportation as a major challenge facing them. Poor road network and high transportation fares increases the transaction costs of transporting raw materials from the producer field to the firms' operating points as well as delivery and distribution to retailers.

Figure 6.5 Major Constraints Faced by Agro-processing Firms in Acquiring Raw Materials

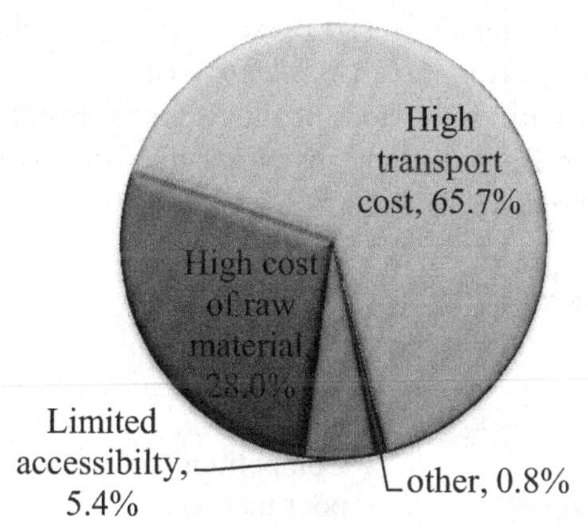

Source: Field Survey 2012

6.8.5 Demand

Low patronage of locally processed foods by consumers is major hurdle to food crop agro-processing firms. Many consumers still prefer unprocessed products due to lack of trust and confidence in the quality of locally processed products. With the advent of many cheap foreign imports, demand for locally processed foods is dwindling. Fluctuating demand affects firms operation through continuous re-adjustment of operational capacity as well as stock scheduling. Large scale food crop agro-processing firms might have to scale back manufacturing capacity or close retail outlets. Also, falling demand affects revenues, which could lead to losses without expense reductions.

6.8.6 Extent of the Market, Competition and Liberalised Trade

The existence of large markets offers the opportunity to produce on a large scale and to benefit from the resultant unit cost reductions. However, most markets for food crop agro-processing firms in Ghana are either small or undeveloped. Also where markets exist for processed food crops, there are several barriers to entry, particularly for micro and very small food crop agro-processing firms, thereby denying them accessibility to the market. As shown in Figure 6.6, about 30% of the firms interviewed indicated lack of accessibility to market as the major marketing constraints. Even among the firms that have overcome this hurdle of entry barriers, there is fierce competition among them as each competes for a market share. The impact of trade liberalization and dumping of cheap foreign imports is also presenting a great challenge to food crop agro-processing firms in Ghana. 25.6% of the sampled firms face acute competition with imported products as a result of trade liberalization. Prices were also in some instances determined by buyers who sometimes buy on credit.

Figure 6.6 Major Marketing Constraints

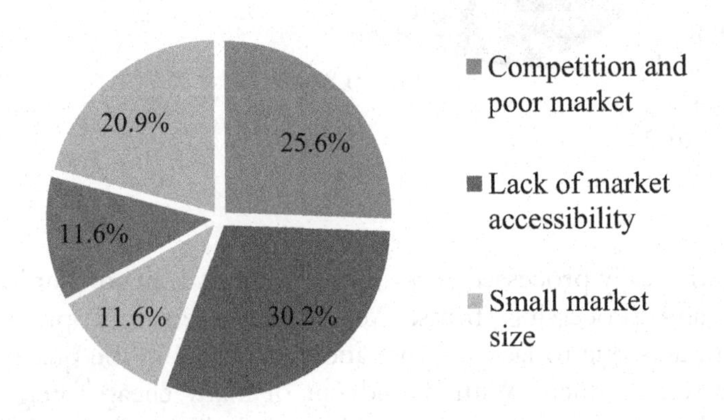

Source: Field Survey 2012

Very few firms are actually using demand generating strategies such as advertising and branding. As shown in Figure 6.7, only 5.1% of the firms interviewed are using advertising to create new market and demand for their products. About 64% and 26% of the food crops agro-processing firms in Ghana use sales representatives and marketing agents respectively. The remaining 5.1% use other methods such as direct marketing.

Figure 6.7 Marketing Strategies of Food Crop Agro-processing Firms

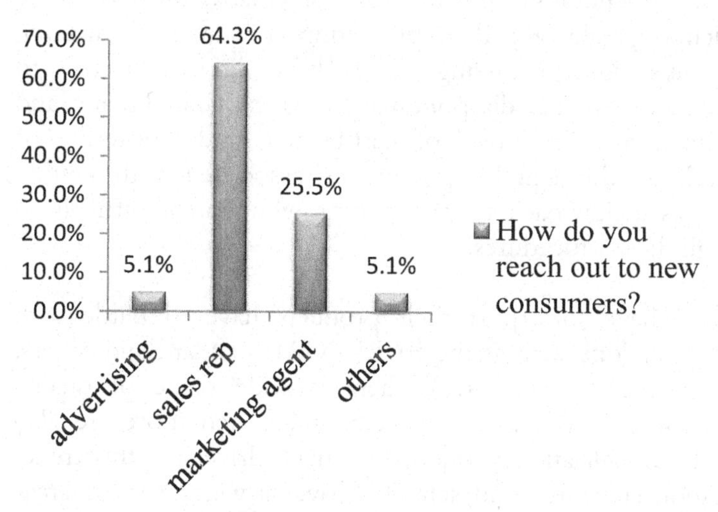

Source: Field Survey 2012

6.8.7 Inadequate Support Services

Inadequate support services from research and training institutions, private sector consultants, small enterprise advisors and engineering workshops also affect food crop agro-processing firms operations. Firms exist to supply consumers' needs. Consumers needs do change over time. Therefore there is the need to continuously innovate to meet changing consumer needs. Most of the firms neither undertake research nor liaise with research institutions and engineers to research into new business areas. Research and training institutions, private sector consultants, small enterprise advisors and engineering workshops will be useful for advancing the improvement of the operations of food crops agro-processing firms. There is also the need to improve the awareness and interests of firms for some of these training workshops.

6.8.8 Strong and Rigid Local and International Certification Procedures

Rigid regulatory and institutional frameworks both at the local and international level pertaining to food safety and hygienic practices which need to be adhered to by food crop agro-processing firms also pose challenges to the operations of the firms in Ghana. Most of these measures which originally aimed at protecting consumers, is making most food crops agro-processing firms have limited access to international markets (such as diaspora markets in the United States and the United Kingdom). The costs of certification and registering of product as well as the lengthy procedures associated with getting certified serves as a drawback to many firms who find it difficult to navigate through these procedures.

Local firms that have to export their products have to battle with phytosanitory regulations such as the EUREPGAP standard and others. According to Heyder *et al.*, (2010), there is wide range of quality certification schemes in operation in recent times and more are also being instituted. Globalization is one of the main drivers to this trend. These certification standards and schemes cover a wide array of areas including production and processing techniques, ingredients or inputs, origin of raw materials, animal welfare, environmental and labor standards as well as compliance with compulsory hygiene and food safety standards.

6.8.9 Macro-economic Policies

Some macro-economic policies such as duty and tax on imported raw material and spare parts, inflation, subsidy and trade openness affect food crop agro-processing operations. According to Dawson (1994) and Simalenga (1996), many policies implemented by governments have served to hinder the development of food crop agro-processing operations. For example, government policies on duty and taxes charged on imported equipment may discourage local food crop agro-processing operations. Firm owners often found it frustrating when they pay high duty and taxes on raw materials and equipment while competing with finished products imported at low duty as a result of trade openness. High inflation also wipes firms' capital gains and reduces the real value of firms' capital and profitability. Changes in government policies also place new burden on food crop agro-processing firms (Dinh et al, 2010). For example, raising payroll taxes could increase operating costs, which could hamper food crop agro-processing firms operations, profitability and growth plans.

6.9 Business Expansion Plans

Every business that starts has the dream of expanding to reach all its prospective customers that may be dispersed in different locations. Business expansion is another means for increasing profitability and extending corporate influence and market share. Business expansion involves opening up new outlets or branches in different locations, increasing product or service offerings. Firms usually fund business expansion through mergers and acquisitions, floating of shares, re-investment or ploughing back profit.

These are funding and business expansion transactions that may be unavailable to several food crops agro-processing firms in Ghana. Business expansion may also require on-boarding of new human resources skills or training, changing or modification of operational process and others. All these must be effectively managed by the firm so that the firm can benefit from such expansion. Thus, business expansion may not necessarily lead to business growth if it is not properly managed. Therefore, businesses need to carefully plan this to achieve the intended purpose. To be able to effectively achieve this, businesses must develop a strategy through which growth can be monitored, maintained and managed.

In the study, food crop agro-processing owners were asked whether they have any plans to expand their business. As indicated in Figure 6.8, about 94% of the firms indicated that they have business expansion plans. The remaining 6% were; however, not optimistic of expanding their business.

Figure 6.8 Food Crop Agro-processing Firms Intension to Expand

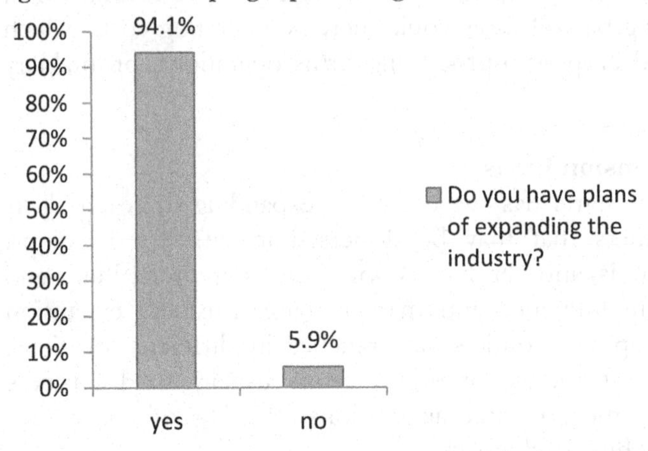

Source: Field Survey 2012

Respondents who indicated their preparedness to expand their business were further asked how they intend to do so. As shown in Figure 6.9, 80% of the respondents plan to expand their business through credit acquisition. Another 10% considered the option of floating shares. Only 2% and 1% respectively, considered re-investment of profit and mergers respectively as their business expansion strategy. This makes availability and accessibility to credit as the major source of funding for business expansion to food crops agro-processing firms in Ghana.

Figure 6.9 Food Crop Agro-processing Firms Expansion Plans

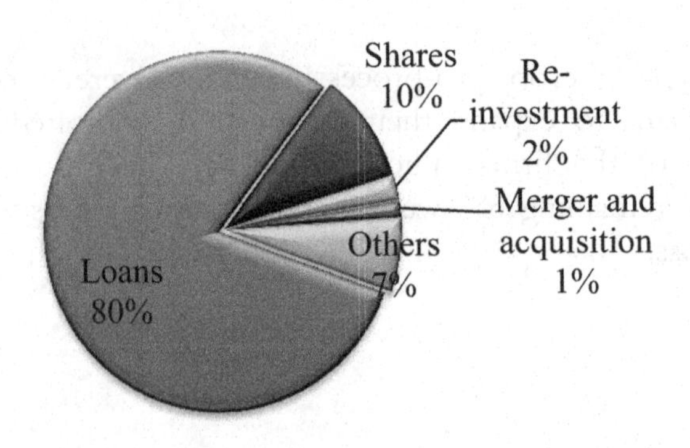

Source: Field Survey 2012

6.10 Conclusions

The potential transformative role that agro-processing firms could play in Ghana has not been realized due to several critical operation constraints facing them. Several of the firms interviewed are operating below recommended capacity with indigenous and inefficient technologies. The study also established that many of these food crop agro-processing firms have no or weak internal and external linkages. They lack adequate human resource skills as well as operating capital to manage the growth and profitability of the firms. Although most of the firms have the intention to expand, their greater reliance on credit as the major source of funding (which is usually unavailable to them) has limited their business expansion.

Financial institutions will have to design firm specific agro-processing loans with low interest rate and longer moratorium to enhance food crop agro-processing firms' access to credit. Government should also provide guarantee to micro, small and other food crop agro-processing firms which lack collateral to be able to access credit. Food crop agro-processing firms should also form associations so that they can benefit from tax exemptions and bargain for a more transparent sanitary and phytosanitory standards. In the interim, food crop agro-processing firms should be sensitized and encouraged to exploit internal and external linkages that has positive impact on firm growth and profitability.

CHAPTER SEVEN: PRODUCTIVITY AND EFFICIENCY ANALYSIS OF FOOD CROPS AGRO-PROCESSING FIRMS.

7.1. Theory and Concept of Production

Production is generally defined as the economic process of converting inputs into outputs. The inputs are the resources used in the production process. Output could be either goods or services by a firm. The inputs used in production include labor, capital and raw-material. In the case of food crop processing, production would be the process of value addition by converting raw food materials from the farm into products with for example, enhanced flavor, taste, texture, shelf life and improved packaging. The rate at which the output of goods and services are produced per unit input is termed productivity. For example labor would be said to be productive if the same number of labor are able to produce larger quantity of goods or the same quantity of goods can be produced by a smaller number of labor (Kipene et. al. 2013). The chapter applies the theory of production and production function estimation in order to analyze the productivity of the inputs used in the food crops agro-processing sector in Ghana.

The relationship that describes how the inputs can be transformed into the maximum attainable output is a production function. The output of a firm is dependent on its labor, capital used, raw materials, entrepreneurial activities and management. In other words, a production function is defined as the maximum amount of output that can be produced (through the use of a given production technology) with a given amount of input (Heathfield, 1971). Different forms of production functions have been widely applied in several studies. They include the linear production function, the fixed-proportions function, the Cobb-Douglas production function, the quadratic production function, the trans-log production function and the constant elasticity of substitution (CES) production function.

In order to analyze the productivity of agribusiness firms in Abia State, Nigeria, Nto and Mbanasor (2001) used the maximum likelihood estimates to estimate a stochastic frontier production function using the specification of the cobb-Douglas function. The variables they employed include labor measured by wage (cost of labor), raw material inputs measured by the cost of raw materials used in the production, capital input defined as amount spent on assets for production and a dummy of firm scale operation. The study found a positive and significant relation between labor (skilled) and total factor productivity. They also found a negative and significant relationship between the cost of raw materials and productivity of agribusiness firms in Abia state. Nimo et. al. (2012) conducted a study to determine the efficiency of resource use in rice production using the Kpong Irrigation Project (KIP) as a case study.

They selected seventy farmers using the simple random sampling technique and used rice output measured by sales, cost of capital operating machine, cost of inputs such as chemicals and paddy rice seeds, and salaries paid to labor for the production functions. Out of three production functions (linear, semi-log, and Cobb Douglas) estimated, the Cobb-Douglass production function gave the best estimate. Their regression results with Ordinary Least Square (OLS) showed that the farmers were in the second stage of production, which is, decreasing returns to scale.

They also estimated the allocative efficiency of resource use and the computation indicated that land (6.63), fertilizer (1.76) and seed (10.84) were being underutilized. Labor (0.000036) and chemicals were, however, being highly over utilized in the study area. Wongnaa and Ofori (2012) also applied the Cobb-Douglas production function to estimate the productivity and resource-use efficiency in cashew production in the Wenchi Municipality of Ghana. They used the OLS technique to regress cashew nut output in kilograms on farm size in acres, labor quantity in man-days, physical capital in Ghana Cedis, liquid fertilizer and pesticides in liters. Their results showed that farm size, capital, fertilizer and pesticides are positively related to cashew output while labor is inversely related. They also found out that farmers were inefficient in the use of resources.

This chapter models the relationship between the output and the productions factors of food crops agro processing sector using the Cobb-Douglas production function. The Cobb-Douglas function is chosen because it has been considered a possible way of characterizing many real-world production processes and often used by economists to study issues related to input productivity. It is also seen as having a robust function, with computational ease and can easily be interpreted (Eze, 2003 and Goni et al, 2007).

Given a simple production function

$$Q = f(x_1, x_2, \dots x_n) \dots \dots \dots \dots \dots \dots \dots \dots \dots \dots \dots \dots \dots \dots \dots (1)$$

Where Q is the output and $x_1 \dots x_n$ are the factor inputs

The general mathematical form of the Cobb-Douglas production function is given by

$$Q = \mathrm{A} \prod_{n=1}^{N} x_n^{\beta_n} \dots \dots \dots \dots \dots \dots \dots \dots \dots \dots \dots \dots \dots \dots \dots \dots \dots \dots (2)$$

Where Q is the output, A is the Total Factor Productivity, x_n represents the inputs β_n represents the elasticities.

To find out the productivity of the input factors of agro-processing firms in Ghana, the functional relationship is specified as

$$Q = f(LAB, CAP, RAW, ENE, TEC) \dots \dots \dots \dots \dots \dots \dots \dots \dots (3)$$

Where,
Q = Output of firms (measured by yearly sales of output in Ghana Cedis)
LAB= Labor (measured by yearly salaries paid to employees in Ghana Cedis)
CAP= Physical Capital (measured by the cost of Machines in Ghana Cedis)
RAW=Raw Materials (measured by the cost of raw materials in Ghana Cedis)
ENE= Energy used (measured by the bill paid on electricity or the cost of fire wood in Ghana Cedis)
TEC=Major Technology used (dummy with 0.5 being manual and 1 being motorized) the dummy is based on the assumption that firms using manual technology are half as productive as those using motorized.

This gives the Cobb Douglas function

$$Q = A\ LAB^{\beta_1}CAP^{\beta_2}RAW^{\beta_3}ENE^{\beta_4}TEC^{\beta_5} \ldots\ldots\ldots\ldots\ldots\ldots\ldots 5)$$

The OLS technique is used to estimate equation (5). The estimation would be also done for different firm scales, firms in different locations and type of crops being processed. In order to estimate using the OLS to satisfy the Classical Linear Regression for a Best Linear unbiased Estimation, the natural logs of the variables are taken on both sides and specified in equation (6) below. The natural logs are also found to transform the variables to a uniform distribution and to improve interpretation.

$$lnQ = lnA + \beta_1 lnLAB + \beta_2 lnCAP + \beta_3 lnRAW + \beta_4 ENE + \beta_5 TEC \ldots\ldots\ldots\ldots\ldots\ldots. (6)$$

The sum of the elasticities i.e. $\sum_{i=1}^{5} \beta_i$ represents the firms' returns to scale.

$$if \sum_{i=1}^{5} \beta_i > 1\ then\ firms\ have\ increasing\ returns\ to\ scale$$

$$if \sum_{i=1}^{5} \beta_i < 1\ then\ firms\ have\ decreasimg\ returns\ to\ scale$$

$$if \sum_{i=1}^{5} \beta_i = 1\ then\ firms\ have\ constant\ returns\ to\ scale$$

7.2 Efficiency Estimation

The resource use efficiency of the various firms is also calculated. This is done according to the various product types since firms have different prices for the different products they produce. To ensure maximum efficiency of resource use, a firm must utilize resources at the level where their marginal value product (MVP) is equal to their marginal factor cost (MFC) under perfect competition. The efficiency of a resource would be determined by the ratio of MVP of the variable inputs (in this case, labor, raw materials and energy) and the MFC. Following Goni et al. (2007), Fasasi (2006) and Stephen et al (2004), the efficiency of resource use is given as:

$$r = \frac{MVP}{MFC} \quad (7)$$

Where,
r = Efficiency coefficient
MVP = Marginal Value Product
MFC = Marginal Factor Cost of inputs

But $MFC = P_x$

Where,
Px = Unit price of input, X.
But $MVP = MPP_X P_Y$

Where, MPP_X is the Marginal physical product of X
And P_Y is the unit price of output.

The Marginal Physical Product (MPP) is given by

$$MPP_X = \beta_x \times APP_x$$

And $APP_x = \frac{\bar{Y}}{\bar{X}}$

Where APP_x is the average physical product

$\bar{Y}$ is the mean value of sales and $\bar{X}$ is the mean value of factor input

β_x is the elasticities or regression co efficient estimated in equation (6)
After substitution equation 7 is written as:

$$r = \frac{\beta_x \times \frac{\bar{Y}}{\bar{X}} \times P_y}{P_x} \quad (8)$$

MVP for each input is therefore obtained by multiplying the regression coefficient of that input with the ratio of the mean value of output and that of input and with the unit price of output obtained from market. MFC of each input is also obtained from the unit market prices of the various inputs.
The decision rule for the efficiency analysis is, if:

$r = 1$; resource is been used efficiently
$r > 1$; resource is underutilized and increased utilization will increase output.
$r < 1$; resource is over utilized and reduction in its usage would lead to maximization of profit.

7.3 Characteristics of Data

Table 1 below presents a description of the data that was employed in the Cobb-Douglas production function. The mean, maximum and minimum values of the raw data are shown in the table. The average sale for the food processing firms in Ghana is US$ 39,425 with the maximum being US$ 6,643,352. The average and minimum cost of raw materials are GHC 8,114.36 and GHC 240.00 respectively. The variables are log transformed and used in the econometric estimation. The description of the variables in skewness and kurtosis are also presented in table 7.1. The skewness is a measure of asymmetry of the distribution of the series around its mean. The skewness of a normal distribution is zero. With the exception of real Cost of energy and major technology used, the other variables show normal distribution.

Table 7.1: Descriptive Statistics of the Data used for the Cobb-Douglas production function

Statistics	Q	LAB	CAP	RAW	ENE	TEC
	GHC	GHC	GHC	GHC	GHC	GHC
Mean	72,994.01	3,013.34	2,560.82	8,114.36	1,201.23	0.54
Max	12,300,000.00	36,960.00	100,000.00	360,000.00	13,440.00	1
Min	840	120	35	240	12	0.5
Skewness	0.2384657	0.789582	0.5390189	0.1090247	-1.737131	3.07435
Kurtosis	3.837395	2.650794	4.068747	2.031514	10.29201	10.4516
Sd	753590.3	5509.667	9717.459	24575.09	1105.196	0.13658
N	272	272	272	272	272	272

Source: Field Survey 2012

Appendix 1 presents the description of the variables by regions, firm scale and crop type.

Table 7.2 presents a pair wise correlation matrix. These correlations show how the variables are inter-related and interdependent. A higher correlation among the explanatory variables may presuppose the presence of multicollinearity. The pair wise correlation matrix shows that, cost of raw materials is correlated with output. This is evident in the values of the correlation matrix which is as high as 0.6334. However, pair-wise correlations may be a sufficient but not a necessary condition for the existence of multicollinearity (Gujarati, 2004).

Table 7.2: Pairwise Correlation matrix for variables

	LnQ	LnLAB	LnCAP	LnRAW	LnENE	LnTEC
LnQ	1					
LnLAB	0.3468	1				
LnCAP	0.2451	0.3499	1			
LnRAW	0.6334	0.3788	0.2103	1		
LnENE	0.1363	0.2668	0.191	0.1856	1	
LnTEC	0.146	0.4409	0.1455	0.2274	0.1232	1

Source: Field Survey 2012

7.4 Econometric estimation and statistical verification of the model

The productivity of the various inputs used by food processing firms in Ghana is derived from the estimation of the Cobb-Douglas production function in equation 6. The estimation was done using the OLS with STATA 11. The coefficients of the Cobb-Douglass production function represents the output elasticities of the various inputs used in food crop agro processing firms. Output elasticity measures the responsiveness of output to a change in levels of the respective inputs, ceteris paribus.

The sum of the elasticities (i.e. the coefficients) gives the returns to scale for the firms. Returns to scale refers to a technical property of production that examines changes in output subsequent to a proportional change in all inputs (where all inputs increase by a constant factor). If the return to scale is equal to 1, then there are constant returns to scale (CRTS), which means output increases by that same proportional change as the inputs. If the return to scale is less than 1 then there are decreasing returns to scale (DRS) implying that output increases by less than that proportional change in inputs. If the returns to scales is greater than 1 then there are increasing returns to scale (IRS) which means output increases by more than the proportional change in inputs.

Table 7.3 presents the results of the estimates of the elasticities of food crop agro processing firms in Ghana. The value of adjusted R^2 is 0.413

which indicates that the model has explained 41.3 percent of total variation in sales of agro-processing firms in Ghana based on the variation in salaries paid to labor, cost of physical capital, cost of raw materials, cost of energy and the type of technology used. The F statistic tells the overall significance of the model. The 1 percent significant level F-statistic indicates that the overall model is significant which means that at least one of the explanatory variables significantly affects the sales of agro-processing firms in Ghana. Apart from the cost of energy and type of technology used, all the other variables are significant. This implies that the significant variables (cost of labor, cost of physical capital and cost of raw materials) are the major contributors of the 41.3 percent (adjustedR2) variations in the dependent variable.

Table 7.3: OLS estimates for coefficients (Agro processing firms in Ghana)

LnQ	**Coef.**	**Standard Error**	**t-ratio**	**Prob**
LnLAB	0.148168**	0.0708691	2.09	0.038
LnCAP	0.111687*	0.0618979	1.8	0.072
LnRAW	0.528669***	0.0460855	11.47	0.000
LnENE	-0.03122	0.1006818	-0.31	0.757
LnTEC	-0.41592	0.4226146	-0.98	0.326
Constant	3.277372***	0.8825354	3.71	0.000
Returns to Scale	0.341381		**Prob > F**	0.000
Number of obs	272		**R-squared**	0.4238
F(5, 266)	39.13		**Adj R-squared**	0.413

Source: Field Survey 2012 * *Significant at 10 percent* ** *significant at 5 percent* ****significant at 1 percent*

The intercept which represents the total factor productivity is significant at 1% level. It indicates the level of output when the value of all independent variables is zero. The coefficient of salaries for labor, cost of physical capital and cost of raw materials are positive and significant at 5 percent, 10 percent and 1 percent respectively. The cost of energy and the type of technology used are negative but not significant. From the estimation, a 1 percent increase in the salaries paid to employees of the agro-processing firms in Ghana would increase sales of output by about 0.15 percent. Similarly a 1 percent increase in the cost of physical capital and cost of raw materials would significantly increase the sales of output of agro-processing firms in Ghana by about 0.11 percent and 0.53 percent respectively. The sum of the partial elasticities gives a value of 0.34 which indicates decreasing returns to scale. Decreasing returns to scale means a proportionate increase in all inputs leads to a less than proportional increase in the output.

A brief description of the regression estimates for agro-processing firms in the various regions are presented in Table 7.4. The F-statistic indicates that the estimation for Eastern region is not statistically significant. This may be due to the limited number of respondents from that Region. The regression estimates for the other regions are however statistically significant at 1 percent. Whilst the R^2 for Brong-Ahafo show that 60.5 percent of the variations in the sales of food crop agro processing firms are explained by the explanatory variables only 29.2 percent of the variations in the sales of output for Central Region firms are explained by explanatory variables. The results show that the cost of raw materials is positive and significant at 1 percent for all the regions apart from Eastern. This implies that cost of raw material is a very important factor for agro-processing firms in Ghana irrespective of the region. Elasticity for cost of physical capital in Central and Western region is significant at 10 percent. Elasticity for labor in terms of salaries paid to workers is significant at 1 percent in Northern and Brong Ahafo Regions. The implication here is that for Central Region, the costs of machines are more important than the salaries paid to labor.

However, cost of labor is a very significant factor for agro-processing firms in the Northern, Brong Ahafo and Western Regions. It is however surprising that a 1 percent increase in the salaries of employees in the Northern region leads to a 0.57 percent decline in the sales of output of food processing firms. This could be due to the predominantly labor-intensive and rudimentary processing technologies that generate minimum value addition. The firms in the Northern region are also characterized by low investment in physical capital. Also the negative marginal productivity of labor may be because of the low skill of the labor that is employed. This could be because the firms' expenditure on salaries is more than the available work for the employees or some employees in the Northern Region are not productive. The results also suggest that whilst Central and Western regions have increasing returns to scale, the Northern and Brong-Ahafo Regions have decreasing returns to scale.

Table 7.4: OLS estimates for coefficients (Agro processing firms of various Regions in Ghana)

LnQ	Eastern	Central	Northern	Brong Ahafo	Western
LnLAB	0.000291	0.035619	-0.573945***	0.3611007***	0.3153055**
LnCAP	-1.62711	0.333248*	0.01635	0.1062	0.1928341*
LnRAW	-0.12253	0.6105317***	0.396406***	0.3662349***	0.360617***
LnENE	0.000659	2.258383	-0.01406	-0.218569	0.016834
LnTEC	-0.00028	0.385166	-0.05177	0.050817	0.913266
Constant	20.16938	-13.6718	9.937355***	4.780932***	3.406228***
Returns to Scale	-1.74896	3.622947	-0.22702	0.665784	1.798856
Number of obs	12	63	41	60	96
F(5, 266)	0.69	6.11	7.04	19.07	14.43
Prob > F	0.6521	0.0001	0.0001	0	0
R-squared	0.3636	0.3491	0.5015	0.6384	0.4449
Adj R-squared	-0.1666	0.292	0.4303	0.605	0.4141

Source: Field Survey 2012

* *Significant at 10 percent* ** *significant at 5 percent* ****significant at 1 percent*

Table 7.5 presents a summary of the regression results for the agro-processing firms in Ghana at different firm scales. Again, because of the low number of respondents for medium scale firms, the regression results are not statistically significant (with an F probability of 0.59) and all the variables are not significant. The R^2 values indicate that the variations in the sales of output for firms producing at micro scale are explained by about 40 percent of the variables whilst for very small-scale and small-scale agro processing firms, about 65 percent and 77 percent respectively of the variations in sales of output are explained by the variables.

It is interesting to note that the cost of raw materials is again positive and significant at 1 percent for the micro, very small and small scale firms. It is seen from the results that as firms increase their scale from micro to very small to small, they effectively use raw materials to enhance productivity thus the small scale firms are seen to be more productive in the use of raw materials since there is a more than 1 percent increase in the sales of output (1.16 percent) after a 1 percent increase in raw materials. There is also a 1 percent and 10 percent positive significant estimate for salaries of labor for micro and small scaled firms. If salaries paid to labor increases by 1 percent for a small scale firm, sales of output increases by 0.93. This indicates that labor is a very important factor for small scale firms. The micro, very small and small scale firms are all experiencing decreasing return to scale. However, return to scale improves when the firm operates at small-scale than at very small and micro-scales.

Table 7.5: OLS estimates for coefficients (Agro processing firms of various Scales in Ghana)

LnQ	Micro	Very Small	Small	Medium
LnLAB	0.2646727***	-0.0298868	0.9271203*	-0.1813
LnCAP	0.0876519	0.032853	0.18514	1.731429
LnRAW	0.4511816***	0.7122916***	1.160781***	0.842009
LnENE	-0.0253692	0.5230225	0.734136	-1.5037
LnTEC	-0.0159544	-0.7988121	-2.162296	1.277411
Constant	3.48865***	-0.3597235	-14.78161	2.070929
Returns to Scale	0.7621826	0.4394682	0.8448813	2.165843
Number of obs	232	18	14	8
F(5, 266)	32.24	7.3	9.51	0.95
Prob > F	0	0.0024	0.0032	0.5855
R-squared	0.4163	0.7525	0.856	0.7031
Adj R-squared	0.4034	0.6494	0.7659	-0.0391

Source: Field Survey 2012

** Significant at 10 percent ** significant at 5 percent ***significant at 1 percent*

Finally, Table 7.6 presents the results for the different firms according to the major products of the food crops agro-processing firms. The regression results for firms processing vegetables and non-alcoholic are not presented because the numbers of observation for them are only 2 and 4. Also, from the F statistics the regression results for Alcoholic, Soap and Confectionary processors are not statistically significant. The regression results for the other categories processing firms are statistically significant. The R-square values indicate that 55.8 percent, 74.4 percent and 37.8 percent of the variations in the sales of output are explained by the variables for root and tuber, cereals and oil and oil-by product processors respectively. Apart from cost of machine all the variables for the oil and oil by-product processors are significant. Cost of energy and the type of technology used are negative for the oil and oil by-product processors. The negative relation between technology and sales mean that at the current market price of oil and oil by-products, sales of output would increase by 2.7 percent if the manual forms of technology are used rather than advanced technology. The use of modern technology may require reasonable increase in market price for the products. The results may also indicate the experience of workers in manual processing and the relatively higher productivity of labor for firms involved in this type of processing (refer to sections 6.3 and 6.4 for the types of activities and the level technology used by firms in this category).

The results once again suggest that the costs of raw materials are important to all firms because they were all significant and positive. The root and tuber and cereal processors also have the cost of capital as positive and significant. A 1 percent increase in the cost of physical capital would increase the sales of output by 0.3 percent and 0.4 for root and tuber and cereal processors respectively. Only oil and oil by-products firms have labor to be significant and positive variable. A 1 percent increase in the salaries paid to labor involved in the processing of oil products would increase the sales of oil products by about 0.3 percent. This again illustrates the important role of labor for firms involved in oil and oil by-products. It is interesting to note from the results presented in table 7.6 that all the firms included in the analysis have increasing returns to scale although the overall regression (Table 7.3) with the excluded firms (due to low sample size) involved in vegetable and non-alcoholic beverages indicate decreasing returns to scale.

The output of a group of firms in the industry may be less than that of the overall and there could be increasing returns at relatively low output levels, and decreasing returns at relatively high output levels.

Table 7.6: OLS estimates for coefficients (Agro firms in Ghana categorized by the crops they process)

LnQ	Root and Tuber	Cereals	Oil and Oil By-product	Soap and Confec-tionary	Alcoholic
LnLAB	-0.05524	-0.17943	0.287854***	0.421694	2.372207
LnCAP	0.313741**	0.404876**	0.03983	-0.05626	-2.20387
LnRAW	0.750914***	0.287071**	0.433445***	0.160556	1.50276*
LnENE	0.090034	0.042049	-0.21866*	-0.66397	-1.98734
LnTEC	0.371828	0.724165	-2.72284***	1.963836	2.005191
Constant	1.430229	6.04751	3.121719**	10.25762*	8.143126
Returns to Scale	1.47128	1.278729	2.18037	1.825857	1.688942
Number of obs	66	24	154	15	7
F(5, 266)	15.12	10.44	17.96	2.36	14.48
Prob > F	0	0.0001	0	0.1247	0.1968
R-squared	0.5575	0.7436	0.3776	0.5669	0.9864
Adj R-squared	0.5206	0.6723	0.3566	0.3264	0.9183

Source: Field Survey 2012

7.5 Efficiency of Utilization of Resources

Table 7.7 presents the resource-use efficiency analysis for firms processing Root and tuber. The major product they produce is 'gari" with an average sales price of US$1.08 per kilos. It is seen from the results that the MPP value for raw materials is higher than for energy and labor. This suggests that if additional raw materials (like cassava) were available, it would lead to an increase in 'gari' production by about 20 kg among the firms. This may have resulted from the fact that the cost of cassava, the major raw material for 'gari', is relatively very cheap because it is one of the major staple crops in Ghana. It also implies that firms in 'gari' processing are more technically efficient in the use of the raw materials. Given the level of technology and prices of both inputs and outputs, the ratios of the MVP to the MFC for 'gari' processors are greater than unity (1) for the inputs except labor. This implies that the inputs are underutilized and firms could be more efficient if they are increased. The possible reason could be that most of the firms are micro and do not have the capacity to spend more especially on raw materials and energy use.

Table 7.7: Resource-use Efficiency analysis for firm processing Root and tuber (into Gari)

Resource	Elasticity	Mean sales	Mean cost	APP	MPP	Py per kg	MVP	MFC	r
Labor	0.055	219318.9	5254.6	41.74	2.31	2	4.61	3.9	1.18
Raw Materials	0.751	219318.9	8247.3	26.59	19.97	2	39.94	18.31	2.18
Energy	0.09	219318.9	1443.1	151.97	13.68	2	27.37	5.45	5.02

Source: Field Survey 2012

Unlike the 'gari' processors, the inefficiency for rice processors is seen in their over utilization of raw materials and energy. This is indicated in Table 7.8 which shows that the efficiency ratios for raw materials and energy are 0.07 and 0.20 respectively. The implication here is that more cost is spent on raw materials and energy compared to the value of their marginal product. The reason could be that, the motorized equipment used in the milling of rice consumes a lot of energy and firm owners may not be conservative in its use. Again, the efficiency of labor for rice processors is more than unity (2.09), and this is an indication of the fact that the laborers are underutilized to an extent that any increase in labor may bring about an increase in sales of rice.

Table 7.8: Resource-use Efficiency analysis for firm processing cereals (rice)

Resource	Elasticity	Mean sales	Mean cost	APP	MPP	Py per kg	MVP	MFC	r
Labor	0.179	26825	1333.4	20.12	-3.61	1	-3.61	1.73	2.09
Raw Materials	0.287	26825	12574	2.13	0.61	1	0.61	8.98	0.07
Energy	0.042	26825	1023.5	26.21	1.1	1	1.1	5.45	0.2

Source: Field Survey 2012

The resource use efficiency analysis for oil products are presented in table 7.9. The results in table 7.9 suggest that whilst labor and energy are being underutilized, raw materials are over utilized. Energy is being used in processing palm oil to an extent that further increase in its use would increase the output. This could be because producers do not have the required knowledge to harness the required energy needed for oil production. For the firms processing oil products to be efficient, they need to increase the use of labor and decrease the use of raw materials.

Table 7.9: Resource-use Efficiency analysis for firm processing oil products (into Palm Oil)

Resource	Elasticity	Mean sales	Mean cost	APP	MPP	Py per gallon	MVP	MFC	r
Labor	0.289	17436.1	2139.4	8.15	2.35	5	11.73	1.77	6.64
Raw Materials	0.433	17436.1	7941.5	2.2	0.95	5	4.76	15.93	0.3
Energy	0.219	17436.1	1165.2	14.96	-3.27	5	16.36	5.45	3

Source: Field Survey 2012

For Soap processing firms, the results indicate that they are efficient in the use of labor with an efficiency ratio value of 1.20. Whilst raw materials and energy are being over utilized but just like oil processing firms, energy is over utilized to an extent that additional use decreases the output (Table 7.10).

Table 7.10: Resource-use Efficiency analysis for firm processing soap

Resource	Elasticity	Mean sales	Mean cost	APP	MPP	Py per piece	MVP	MFC	r
Labor	0.4216941	7331.21	1484.80	4.94	2.08	0.50	1.04	0.87	1.20
Raw Materials	0.1605563	7331.21	3556.01	2.06	0.33	0.50	0.17	4.17	0.04
Energy	0.6639659	7331.21	753.78	9.73	6.46	0.50	3.23	5.45	0.59

Source: Field Survey 2012

Alcohol processing firms are efficient in the use of labor and raw material however they are inefficient in using energy (Table 7.11). The efficiency analysis suggests that with the current pricing, the only input that could be efficiently used without decline in sales output is labor. Pricing may be related to products quality and firm competitiveness. The next section discusses the concept of product quality.

Table 7.11: Resource-use Efficiency analysis for firm processing Alcohol (Akpeteshie)

Alcoholic beverage- Akpeteshie per litter									
Resource	Elasticity	Mean sales	Mean cost	APP	MPP	py	MVP	MFC	r
Labor	2.372207	271380.02	7617.14	35.63	84.52	3.30	278.90	4.01	69.55
Raw Materials	1.50276	271380.02	7264.02	37.36	56.14	3.30	185.10	20.53	9.02
Energy	1.987344	271380.02	1212.13	223.89	444.94	3.30	1466.97	5.45	269.27

Source: Field Survey 2012

7.6 Conclusion

The chapter analyses the productivity and efficiency of agro-processing firms in Ghana. Different regression results are also estimated for the firms in the various regions, their scale of production and the type of crops they produce. The Cobb-Douglas production function is estimated using the ordinary least square technique and the various returns to scale are determined. The production function has sales of output depending on salaries of labor, cost physical capital, cost of raw materials, cost of energy and type of major technology used.

The results from the study suggest that cost of labor, cost of physical capital and cost of raw materials are significant factors to sales output for food crops agro-processing firms in Ghana. Cost of raw materials is an important factor explaining sales of output irrespective of the location of the firm, type of crop processed and the scale of production. The cost of labor is also an important factor for firms in the Northern and Brong Ahafo regions. It is also a significant factor for firms producing at micro and small scale level and firms who process oil and oil by products. The results indicates that due to some factors such as low skill of labor, under-employment and the predominance of labor-intensive and rudimentary processing technology used particularly in the Northern Region there is negative marginal productivity of labor for food crops agro-processing firms in this region. There will be the need for training and other capacity enhancement activities that will improve the skills and capacities of the workforce involved in food crops agro-processing.

The results indicate that productivity and efficiency of utilizing raw materials improves consistent with scale of operation. As firms expand from micro, very small to small scale, they become more diversified in their products delivery which also improves their efficiency of conversion of raw materials. For example, diversification by roots and tubers into gari-grades, 'agbelima' and 'kokonte'. The sum of the partial elasticities gives a value of 0.34 which indicates decreasing returns to scale. Decreasing returns to scale means a proportionate increase in all inputs leads to a less than proportional increase in the output.

The returns to scale are increasing for firms in their individual processing categories but when assessed in terms of region and scale level, the returns to scale tend to decrease. It is therefore recommended that, government and stakeholders should increase subsidies on inputs given to firm owners which would in the long run reduce the cost of raw materials for agro-firms. Interventions such as credit facilities and trainings on financial and human resource management should be given to agro-processing firms based on the type of crop being processed since the productivity for the different crop categories exhibit increasing returns to scale.

The analysis also shows a considerable room for improvement in the productivity and efficiency of food crop processing firms in Ghana. The results of this study suggest that, firms could increase sales of output through better use of available resources. The achievement of efficiency in all the inputs and for that matter their total productivity will depend on increasing all the inputs that need to be increased, and reducing all those that need to be decreased. Firms need some training on efficient allocation of funds to the various inputs that are being used.

Although there are standards for food processing in the country, there is the need for enforcement, education and involvement of food processors in policy and standard formulation and implementation, as well as the need for more bye–laws at the local government level to curb such occurrences.

CHAPTER EIGHT: STANDARDIZATION AND MARKETING OF PROCESSED FOOD PRODUCTS

8.1 Standardization

Standardization of food products refers to the process of establishing uniform technical specifications recognized by applicable scientific institutions and generally accepted in the market. Standards are the means or criteria used to judge products, processes and producers (Hatanaka et al., 2005). Standards assure quality of the products. These standards are often specified as rules or regulations. This is an important pre-marketing and commercialization requirement aimed at assuring product quality and providing the opportunity for firms to use compliance to standardization to increase their competitive edge.

Globally, there are a number of regulations or standards that govern the food industry. Common among these international and trade bloc standards are the Codex Standards established in 1962 by the United Nations Food and Agriculture Organization (FAO) in collaboration with the World Health Organization (WHO), the Agreement on the Application of Sanitary and Phytosanitary (SPS) Measures and the Agreement on Technical Barriers to Trade (TBT) established by the World Trade Organization (WTO) in 1995. Others are the EUREGAP standards, US Food Standards and Food Safety Modernization Act (FSMA) and WHO International Digest of Health Legislation (Hooker and Caswell, 1999; APO, 2005).

Standards can also be set at the national level by private and public organizations (Theuvsen and Plumeyer, 2008; Jahn et al., 2003). According to Theuvsen and Plumeyer, (2008), public standards are laid down by national or regional public institutions such as the Ghana Standards Authority. Private standards are often instituted by non-public institutions such as customers' advocacy groups and organizations, suppliers' organizations and voluntary standardization institutions such as the International Organization for Standardization (ISO).

Product standardization may lead to economic, technical, social, political and management benefits, although not all firms benefit from standardization. Firms may also benefit disproportionately. According to Kotler (2007), product adaption and cultural-specificity may hamper some firms' competitive edge in a standardization effort. Other factors such as diversity in tastes and preferences, income levels, and perceptions may influence people's tolerance of risk and requirements for standardization. For instance, Coca-Cola has to withdraw its 2-liter bottle in Spain after discovering that few local refrigerators had large enough compartments. Also, the combined effects of institutional weaknesses and rising compliance costs could contribute to the further marginalization of weaker economic players, including micro, very small and small food crop agro-processing firms. Standards may also 'raise the bar' for new entrants. These notwithstanding, standardization remain an important first step in assuring consumers of quality products and maintaining their confidence

Quote 8.1

FRUITS BAN: Agric Ministry admits Ghana can't meet all US standards

Fruits

The Ministry of Agriculture says Ghana may have to continue exporting fruit to the European market because it simply cannot meet all the conditions set by the United States of America.

The US has banned the export of fruits from Ghana and 27 other African countries for poor quality standards.

Source: Myjoyonline.com, 2013

and trust in a product. Before discussing measures for assurance and product quality of food crops agro-processing firms and how it is affecting their competitiveness, it will be appropriate to discuss what a product quality is. This is because the quality of a product is a debatable concept that has been defined differently from the perspective of consumers, producers, traders, economist and philosophers. As illustrated in the quote 8.1, the USA banned fruits from Ghana because they could not meet the USA standards requirement (Myjoyonline, 2013). This has strong implications for market opportunity, premium pricing for organic fruits by "green" consumers that are predominantly located in the USA.

8.2 Concept of Product Quality

Globally, food crop producers, processors and consumers always try to pursue higher quality products. However, the term 'quality' is often unclear as its meaning differs for different people or under different circumstances and from one geographical setting to the other. Generally, five broad approaches for explaining product quality can be identified in the literature. These are the (1) transcendent approach of philosophy, (2) the manufacturing-based and (3) value-based approach of marketing and operations management, and the (4) product-based and (5) user-based approach of economics. The transcendent approach borrows from the Platonic ideologies by arguing that quality is an innate excellence which users recognize through experience.

The manufacturing-based approach defines quality as a conformance to standards or specifications. Any deviation from this standard thus implies reduction in product quality. The Value-based approach defines quality in terms of prices or cost. Thus a quality product is one that provides performance at an acceptable price or conformance at an acceptable cost. Contrary to the value-based approach, the Product-based approach defines quality product as one that has the highest amount of desired ingredients or attributes. The User-based approach is based on the consumer behavior theory. It assumes that quality lies in the eyes of the beholder, and that quality products are the goods that best satisfy consumers (Edwards, 1968, Garvin, 1988).

However, embedded in each of these approaches is a description of optimum or desirable characteristics of the product. The nextsections therefore describe product quality from the economic and marketing point of view. Thus, from the economic and marketing viewpoint, the quality of foods or products is a composite term encompassing many desirable characteristics or attributes of foods. These desirable product characteristics can be measured in different ways but one popular method is to describe the generally accepted 'quality attributes' of the product.

8.3 Quality Attributes

Quality attributes include any attribute of a product from which consumers derive utility or disutility (Senauer, 2001). Although measures of product quality may vary from one geographical setting to the other, it generally may include appearance, taste, and convenience of preparation, packaging, color, odour or flavor, nutritional value, adulterants and contaminants. It may also include food safety attributes such as absence of foodborne pathogens, heavy metals, pesticide residues, food additives, naturally occurring toxins, fat content, or lack of calories, fiber, sodium, vitamins, and minerals. These quality attributes can be categorized into sensory, health, process or convenience-related attributes (Grunert, 2003).

Sensory attributes relate to the food quality attributes that are generally determined through the sensors such as taste, appearance, texture and smell. Health attributes are increasingly gaining grounds. Increasing education on health and safety are major contributory factors. Consumers now have adequate knowledge of the relationship between health and the food they consume. Therefore consumers use some cues such as the source of product, additives, fat content and others (Brunsø, Fjord, and Grunert, 2002) to determine the quality of the product. These quality attributes are currently being exploited by food crop agro-processing firms in the production of functional foods (Frewer et al., 2003) such as low sugar fruit juice enriched with iron or calcium and other healthy ingredients, high fibrous food product and others.

Process attributes refer to non-market attributes related to the processes along the product value-chain. These include naturalness as in whether the product is organic or not; whether products were produced with due concern for equitable income distribution, environmental considerations; child labor and genetically modified products.

Convenience-related quality attributes basically relate to food products that save time and the energy consumers typically spend on shopping, food storage, food preparation, eating, and food disposal. Convenience attributes encompass the size dimensions, extent of processing required before product can be consumed or stored, storability as well as movability of the product.

Quality attributes, such as brand, price, convenience, freshness and sensory characteristics can be experienced at the time of consumption or purchase. Credence quality attributes, such as healthiness, naturalness and wholesomeness on the other hand cannot be experienced directly. These credence attributes take a relatively longer period for it to be experienced and cannot be experienced at the time of purchase (Wirth et al., 2011). Consumers use a number of intrinsic and extrinsic quality cues at the point of purchase of product to determine the quality of the product. Extrinsic quality cues include price, taste, smell or flavor, texture, freshness, brand name, stamp of quality, country of origin, mode of storage, packaging, production and nutritional information. Intrinsic cues include visible fat, color, sugar content, caffeine and others.

8.4 Quality of Processed Foods From Food Crop Agro-processing Operations

Harvested food crops often need further processing such as cleaning, grading, drying, milling and packaging to get rid of various undesirable materials such as inert particulates, seeds of noxious weeds, other crop variety seeds, decorticated seeds, damaged seed and/or off-size seeds. Processing is also meant to improve preservation and improve the quality of the produce. Ordinarily, some food crops need some form of preparation and processing to make them more attractive to eat. For example cereals, roots and tuber crops, and vegetables are mainly unpalatable in their raw state whiles some food crops such as cassava, are dangerous if eaten without processing (FAO, 2010). Other food crops such as nuts and fruits that are eaten raw could also be processed into a wide variety of other products.

Agro-processing operations have been used to improve the sensory, convenience and health attributes of food since man's existence (Shahidi *et al.*, 2003; FAO, 2010). Therefore one cannot improve the usability and quality of the harvested food crops without going through food crop agro-processing operations. In recent times, growing consumerism across many countries particularly in developed countries has increased the drive for quality in agro-processing operations (Ponte and Gibbon, 2005). Furthermore, consumers have also become more sophisticated and cautious in their consumption of food products due to recurrent food crises (Vermeulen and Bienabe, 2007).

According to Senauer (2001), consumers in Europe and the United States have moved up the Maslow's hierarchy of needs pyramid from satisfying basic physiological needs, including taste and palatability to meeting other non-market needs. The author adds that the traditional demand for quantity as a means of increasing utility has changed with consumers now preferring quality products. Therefore income elasticity in terms of quantity is low whiles the elasticity for food quality attributes, such as nutrition and health, safety, convenience, and diversity, are quite high. The demand for high quality food products has intensified consumer advocacy for improved quality assurance, safety and more stringent standardization (Jaffee and Henson, 2001; Acksoy, 2005). For many high-value foods, the challenges of consumer preference and firm competitiveness have now moved well beyond price, quantity and basic quality to safety and health concerns of products (Buzby, 2003).

Besides what consumers regard as "quality" has changed considerably in recent years. Today, product quality has evolved from the narrowly perceived sensory attributes to include health, process, and convenience attributes (Grunert, 2003). Therefore, the quality of products from food crop agro-processing operations has become an important competitive issue in the food crop agro-processing industry today. Food crop agro-processing firms that fail to improve or maintain the quality of their products therefore become uncompetitive (Gbedemah, 2004). Robinson et. al. (2012) notes the declining competitiveness of Ghana's processed products to imported processed food products due to their relatively low quality and packaging.

Figure 8.1 A young man fetching oil with his hands which is a non-conformity.

Source: Field Survey 2012

8.5 Certification of Processed Food Products

To assure quality of products, it is often necessary to analyze and certify food products before or after processing. Certification refers to the verification of compliance with set standards. Product certification involves the issuance of a certificate or mark or both to demonstrate that a specific product meets a defined set of requirements for that product, usually specified in a standard. The certification mark is normally found on the product packaging (label) and may also appear on the certificate issued by the certification body (GSA, 2013). The mark carries a reference number or name of the relevant product standard against which the product has been certified (GSB, 2011). Product certification ensures that quality assurance is maintained throughout the unit operations of food crop agro-processing firms. Certification provides a means for informing ultimate customers about products, and offers food crop agro-processing firms certificate, a mark or some other signs as a guarantee that the information provided on the product and the production processes are verified by the certification agency. Thus, certification helps to promote consumer confidence in products. Certification also aids customers and other suppliers to identify those food crop agro-processing firms who have invested in production of high quality products; thereby, reducing quality uncertainties in food supply chains (Gawron and Theuvsen, 2009).

In the last decade, many quality assurance schemes have been introduced especially by the private sector (Albersmeier et al., 2010: Sodano, 2006: Jahn et al., 2005) with standards often higher than those of public regulations. Food crop agro-processing firms meeting these standards are often issued with a mark, label or logo through a certification process. It is often not sufficient to just adopt any quality assurance scheme and standards without the appropriate certification and labeling schemes in place.

The benefits of adopting quality assurance systems or certification standards depend on the type of products the firm produces, the kind of market they operate in as well as the certifying body involved (Karipidis et al., 2008). Theuvsen and Plumeyer (2008) indicates that a benefit of certification is to reduce transaction cost by avoiding or decreasing promotional and advertising costs that are usually associated with traditional marketing practices. Certification can also be used to achieve differentiation in products and markets. However, certification increases the transaction costs in the national and international markets and may reduce the scope and profitability of firms (Theuvsen and Plumeyer, 2008).

In Ghana, certification is carried out by the Ghana Standards Authority (GSA) (Aryeetey and Ahene, 2005). However, section 27 of the Food and Drugs Act, 1992 (PNDCL 305B) mandates the Food and Drugs Board to regulate the manufacture, importation, exportation, advertisement, distribution and use of food products on the Ghanaian market with respect to their safety and quality. The Food and drugs Act also describes standards for processed food as well as offenses. For example sections 3, 4, 5, 6, 7, and 8 touches on deception of consumers, standards of foods, prohibition against sale of poor quality food, unwholesome food, sale of food under unsanitary conditions, and food unfit for human consumption respectively.

Yet there have been many crises and scandals in the food crop agro-processing sector minimizing consumer confidence and trust in these regulations. For example, according to a study conducted by the Sunyani Co-operative Food Crop Processors Society Limited as reported by the Ghana News Agency (2013), non-food grade metal materials are used in fabricating 97 per cent of small–scale food processing equipment leading to heavy metals such as iron, cadmium and lead in processed foods. There are also problems with processing food under unhygienic conditions that leads to contamination, food poisoning and other health hazards.

They also found that 95 per cent of food processors did not have a working relationship with the Food and Drugs Board and Ghana Standards Board or did not know the procedures for certifying their products. Although there are standards for food processing in the country, there is the need for enforcement, education and involvement of food processors in policy and standard formulation and implementation, as well as the need for more bye–laws at the local government level to curb such occurrences.

8.6 Rate of Product Quality by Food crops Agro-processing Firms
With the current proliferation of standards and certification schemes, and the advent of consumer protectionism as well as the enactment and rectification of domestic and international food safety laws, the quality of processed food products in Ghana should be improved. The study therefore asked managers to subjectively rate the quality of their own processed food product relative to other same products from other industry players. As shown in Figure 8. 2, 26.8% were not able to rate the quality of their food product whiles 2.2% of the firms self-rated their product as fair. Also, 49.3% and 21.7% self-rated their product as being good and excellent respectively.

Although the result seems subjective, the fact that some managers of food crop agro-processing firms were not confident or were not able to rate the quality of their product shows a clear lack of market research by the firms. This has implications on the marketing of their products. If firm owners cannot affirm the quality of their product, then it brings to doubt the quality of the product. It also raises the question of how they advertise their product or what they tell their consumers or how they convince their potential consumers to increase demand for their product. In essence, unless the firms are monopolistic, the quality of the product should be one of the major factors determining their market share or competitiveness.

Figure 8. 2: Self-rated Quality of Products

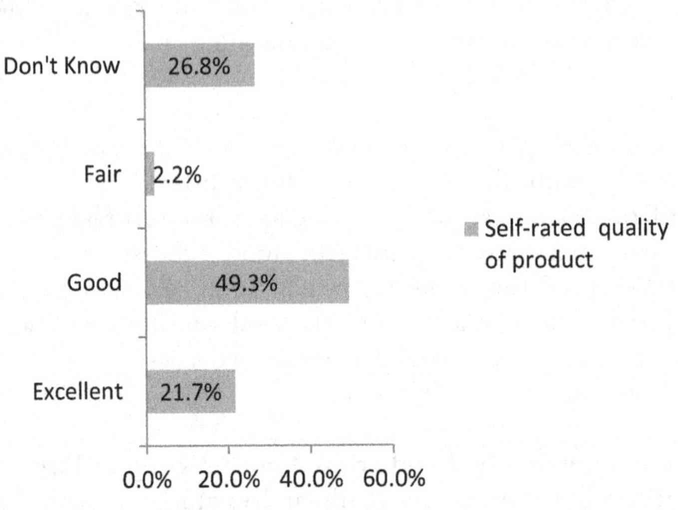

Source: Field Survey 2012

For firms in the industry to remain competitive, they must strive to improve the quality of their products. But improvement in the quality of processed foods will also mean that implementing agencies like the Food and Drugs Board and others are adequately resourced to carry out their mandate.

8. 7 Marketing Opportunities and Strategies

Food crop agro-processing firms, just like any other firms, operate with the aim of generating enough revenue to offset cost. As shown in Figure 8.3, in order for firms to fulfill their profit maximizing objective, they always look for opportunities in the market either by developing new market, increasing penetration, developing new products or diversifying their product mix (Figure 8.3). Thus, firms may sell their current products through increased market penetration or develop new market for their products or may diversify their products mix. Firms could also innovate and develop entirely new products to gain entry into new markets.

Figure 8.3: Typical Market Opportunities and Strategies for Food Crop Agro-processing Firms

	Current products	New products
Current market	Market penetration	Product development
New market	Market development	Diversification

Firms also develop marketing strategies about the specific customers to target and the marketing mix to adopt in order to reach out to new customers or ensure that old customers remain loyal to their products. There are a number of combinations of these marketing mix decisions or strategies and targets that a firm might use. These strategies have generally been classified as direct or indirect marketing strategies. Direct marketing strategies involve product exchanges between the final users of the products (consumers) and processors without intermediaries. Indirect marketing strategies, on the other hand, require use of intermediaries such as marketing agents or sales representatives and others. From Figure 8.4, 64.3% of the food crop agro-processing firms use sales representatives as a marketing strategy. Also, 25.5% uses marketing agents as a means for penetrating, developing new markets or maintaining customer loyalty. Another 5.1% uses advertising whiles the remaining 5.1% uses direct marketing strategies to reach out to customers

Figure 8.4: Marketing Strategies of Food Crop Agro-processing Firms in Ghana

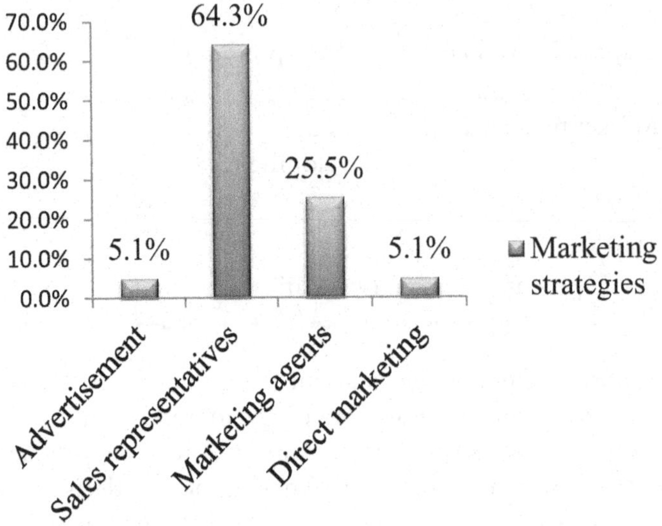

70.0%
60.0%
50.0%
40.0%
30.0%
20.0%
10.0%
0.0%

5.1%
64.3%
25.5%
5.1%

Marketing strategies

Advertisement
Sales representatives
Marketing agents
Direct marketing

Source: Field Survey 2012

8.8 Modes of Selling

Firms employ different selling modes based on the characteristics of their consumers, the nature of their products, the size of their market share, their operational schedules or cycles as well their working capital. Firms whose products are highly perishable and have low market share or low demand normally sell their products on credit basis. Firms whose products are in high demand normally sell their products on cash only basis. Other firms may operate between these two extreme selling modes by combining credit sales with direct payment methods. From Figure 8.5, about 26.8% of the food crop agro-processing firms use credit sales as a mode of selling their processed food products whiles about 45.6% use direct payment method. The rest (27.6%) combines both methods.

Figure 8.5: Modes of Selling by Food Crop Agro processing Firms

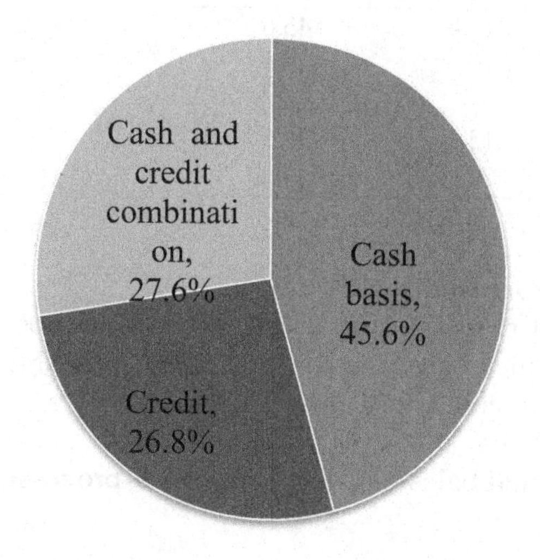

Source: Field Survey 2012

8.9 Credit Policy and Duration of Receivables (Credit Sales)

Although over 27% of the food crop agro-processing firms undertake credit sales, most of them had no clear cut credit policy. None of the firm owners was able to state their policy on the acceptable proportion of the firms product agreed to be sold through credit or acceptable amount of firms' revenue that is to remain as receivables at specified periods. Again, almost all of the firms under study use mere "word of mouth" or no formal agreement to sell their products on credit.

Also, close to half of the firms interviewed have no firm policy on payment durations for their customers. About 45.6 % as well as 3.7% could confirm that they give out their product on credit for a period of less than one month or a month respectively. Food crops agro-processing firms whose credit sales payment duration lasted one to three months were only 2.2%. Firms with short payment durations may have high demand than those with relatively longer payment duration.

Table 8.1: Duration of Credit Receivable from Credit Sales

Duration of payment	Frequency	Percent
Less than one months	124	45.6
One month	10	3.7
one –three months	6	2.2
None	132	48.5
Total	272	100.0

Source: Field Survey 2012

8.10 Annual Sale of Processed Food Crops

From the survey, food crop agro-processing firms could not state the number of customers they have; but most of them were able to state their sales, and this is presented on the bar chart below.

Figure 8.6: Mean Annual Sales of Food Crop Agro-processing Firms.

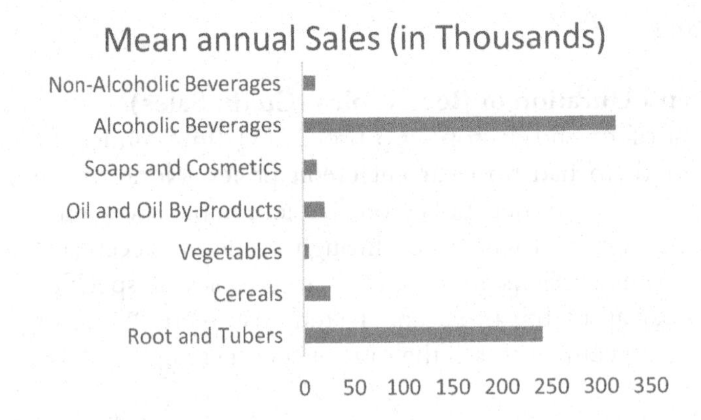

Source: Field Survey 2012

From the chart, food crop agro-processing firms producing alcoholic beverages enjoy the largest annual sales. Firms processing roots and tubers made the second highest sales per annum. Vegetables, soaps and cosmetics and non-alcoholic beverage processors were firms making the lowest sales per annum. Although, there were no available data to confirm that the pattern of sales observed were entirely due to marketing, it could be agreed that increased marketing has the potential to raise the sales of food crop agro-processing firms.

8.11 Some Marketing Constraints

Food crop agro-processing firms may be confronted with several marketing constraints. These constraints may be endogenous especially related to capacity, whiles others are external factors. One of the endogenous marketing constraints is lack of market research. Majority of the marketing constraints firms face are from external sources. These marketing constraints are discussed under various subthemes below.

Lack of Market Research

Market research enhances a firm's capacity to produce new products, develop or differentiate products that meet some segments of the consumer expectations. Market research also helps the firm to develop new market and increase firm penetration and market share. Surprisingly, only 11.4% of the firms interviewed have research department. The remaining 88.6% have no research department. Even those with research department, most of these have not been adequately resourced and hence are not vibrant.

Figure 8.7: Proportion of Firms with Research Department

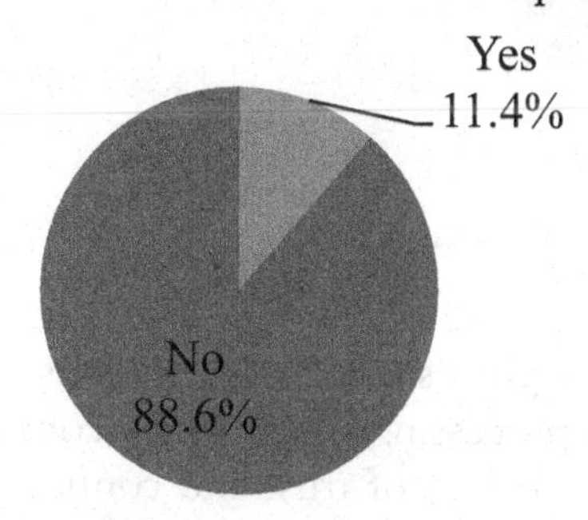

Source: Field Survey 2012

Extent of Market Competition and Liberalized Trade Effects

The existence of large markets offers the opportunity to produce on a large scale and to benefit from the resultant unit cost reductions. However, most markets for food crop agro-processing firms in Ghana are either small and underdeveloped or non-existent. Also where markets exist for processed food crops there are several barriers to entry, particularly for micro and very small food crop agro-processing firms, thereby limiting their accessibility to the market. As shown in Figure 6.6, about 30% of the firms interviewed indicated lack of accessibility to market as the major marketing constraints. Among the firms that have overcome this hurdle of entry barriers, there is still fierce competition among them as each competes for a market share. The impact of trade liberalization and dumping of cheap foreign imports is also presenting a great challenge to food crop agro-processing firms in Ghana. As shown in Figure 8.8, 25.6% of the sampled firms face acute competition with imported products as a result of trade liberalization. Prices were also in some instances determined by buyers who sometimes buy on credit.

Figure 8.8: Constraints Faced by Food Crop Agro-processing Firms in Ghana

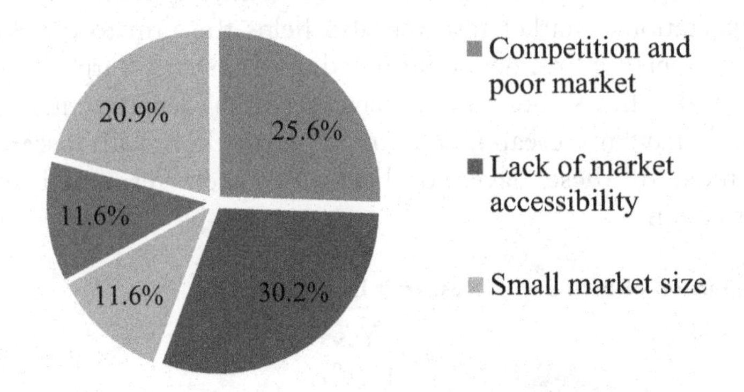

Source: Field Survey 2012

Low Demand

Low patronage of locally processed foods by consumers is a major hurdle to food crop agro-processing firms. Many consumers still prefer unprocessed products due to lack of trust and confidence in the quality of locally processed products. With the advent of many cheap foreign imports, demand for locally processed foods is dwindling. Fluctuating demand affects firms operation through continuous re-adjustment of operational capacity as well as stock scheduling. Large scale food crop agro-processing firms might have to scale back manufacturing capacity or close retail outlets. Also, falling demand affects revenues, which could lead to losses without expense reductions.

Inadequate Support Services

Inadequate support services from research and training institutions, private sector consultants, small enterprise advisors and engineering workshops also affect food crop agro-processing firms operations. Firms exist to supply consumers' needs. Consumers needs do change over time. Therefore, there is the need to continuously innovate to meet changing consumer needs. Most of the firms neither undertake research nor liaise with research institutions and engineers to research into new business areas. Research and training institutions, private sector consultants, small enterprise advisors and engineering workshops will be useful for improving the operations of food crops agro-processing firms. There is also the need to improve the awareness and interests of firms for some of these training workshops.

8.12 Conclusions

The study has highlighted the gains from trade in processed food crops as well as the constraints facing firms in the food crop agro-processing industry regarding standardization, quality control and marketing. It was noted from the study that although the GSA has standards for food processing, most of the firms are not in compliance. There will be the need for the GSA and other relevant institutions to enhance awareness and compliance to these standards in order to improve the quality and competitiveness of processed agro products from Ghana. The enforcement mechanisms should have complementary education and capacity development activities to help the food crops agro-processing firms apply the knowledge to improve the quality of their products. There could also be award and other incentive systems with enhanced quality control and compliance with the standards as some of the major criteria for winning the award and recognition.

The measures for improved compliance to the standards and quality control should form the pivotal components of a strategic national competitiveness program for food crops agro-processing firms. To be able to compete with imported food products, this national competitiveness program will be critical. At least it could focus on the short-term objective of import substitution and then build on that achievement for competitiveness in the international market. If food crops agro-processing firms cannot invest in getting international certification to improve their competitiveness in the international market, at least they could focus on improving quality control to enhance their competitiveness in the national market. Import substitution based on food crops agro-processing is achievable with commitment and support from the government. With the tremendous potential for rural income and employment, the country could use agro-processing to galvanize inclusive growth in the country. Without this strategy, most of the food crops agro-processing firms will not be able to compete with imported products and they will subsequently exit the industry. This will worsen the current high unemployment, food insecurity and poverty in the nation.

The research findings also indicate that for food crops agro-processing firms to improve their competitiveness, they will have to enhance their capacity for packaging, promoting and marketing their products. They will also have to set in place formal credit agreements with their clients and have effective systems for managing account receivables. These business practices are necessary if food crop agro-processing firms are to improve their formality and attractiveness for credit and other financial products to expand their business operations. The perception of their informality is one of the major reasons why financial institutions regard them as high risk group.

CHAPTER NINE: CAPITAL INVESTMENT OF FOOD CROPS AGRO-PROCESSING FIRMS IN GHANA

9.1 Introduction

The low output of food crops agro-processing firms may not be entirely due to lack of competent human resources but may be as a result of key assets. Therefore, to continuously expand and meet consumer demand and expectations, food crop agro-processing firms invest in useful resources that have economic value and can produce cash flows and revenues for the owners. Most food crop agro-processing firms thus devote large amounts of their finances to assets acquisition for their businesses in order to achieve their organizational goals. These may involve investments in capital or fixed assets such as processing or milling plants, boilers, dryers or conveyors and others. Assets are therefore a steady and secure source of earnings and an essential component of every business operation. Lack of key productive assets may undermine the quality and quantity of products required by customers. This may affect customer satisfaction and competitiveness which will consequently hinder the firm's prospect for growth. It is therefore vital for every company to invest and manage their assets properly. Mismanaging these assets could result in problems such as breakdowns, frequent loss of business opportunities, payment of higher assets and property insurance premiums.

This chapter discusses capital investments of food crop agro-processing firms in Ghana. Capital investment connotes financial acquisitions or outlays related to fixed capital or assets. Serious attention must be given to capital investment since it may involve large investment or expenditure on assets with long-term operational life and sometimes long-term loan repayment periods. Firm's capital investment often results in increased capital or fixed assets in the balance sheet or increased liabilities, especially long term liabilities. The extent of capital investment by food crop agro-processing firms can be measured from several analytical options. For the purposes of this study, capital or fixed assets refer to tangible non-current assets that have extended period of use beyond one year and that are not readily convertible to cash. Discussions on current assets and liabilities of food crop agro-processing firms have also been discussed to a limited extent.

9.2 Assets

Assets refer to all manner of resources used in any business operations. These may entail those that are liquid or current as well as those that are not easily sold in the regular course of business's operations for cash. The latter is normally referred to as capital, fixed or non-current assets. These assets combinations give a snapshot of the financial position of the firm. Current assets of food crop agro-processing firms usually include inventory or stock, receivables and cash. Fixed assets are mainly land, buildings, machinery, boilers, conveyors, autoclave or ovens and others.

9.2.1 Current Assets

Current asset refers to cash and cash equivalent assets such as inventories or stock, receivables and others. Current asset is the general name given to all assets that are easily convertible to cash in the normal course of business operation or have recurrent uses. Food crop agro-processing firms have key responsibilities in the management of their cash and cash equivalent assets, if they are to win investor confidence and trust as well as maintain or improve firm's position in the industry.

9.2.1.1 Cash

Surprisingly, most of the firms interviewed had limited record of cash flows. One major reason is that these firms barely keep cashbooks. Also, a good number of the firms had no firm policy on cash that should be held by the firms at a time. This implies poor cash management. Particularly for most micro, very small and small scale food crop agro-processing firms, the tendency to keep much of the firms operating capital as cash on hand not as a deposit in the bank was prevalent. Majority of the firms interviewed have poor financial record keeping skills and inefficient cash management systems.

9.2.1.2 Inventory

Food crop agro-processing firms often hold stocks of raw materials or products. Inventory helps firms to quickly respond to any sudden surge in consumer demand or unexpected sales orders. However, excess or large volumes of inventory could lock a firm's operating capital and could create additional burden in terms of storage space and expenses. On the contrary, low inventory also leads to loss of customers during equipment failure and loss of profit.

Therefore, the volume of inventory to be kept by the firm must be properly planned and executed. Inventory planning is a relevant aspect of food crop agro-processing firms operations, although this is often neglected. Whiles some firms deliberately keep inventory, others may have no inventory due to poor production and sales planning, equipment failure or breakdown or simply from poor firm's judgment. The production and marketing units of the firm have key roles to play in managing inventories. Thus, marketing department must inform the production department how much of the processed products were distributed or sold, how much is left as stock and how much extra should be produced and when to produce these. Firms' inventories must follow industry trends or patterns and must be based on outputs of market surveys.

Unfortunately, since most food crop agro-processing firms have non-functional or no marketing department, inventory scheduling was poor. Whiles some firms had no inventory, others had it in excess of their capacity. From Figure 9.1, more than 45% of the firms interviewed did not have any inventory of raw materials. Also, 34.6% had stock of raw materials that could last for a week whiles another 1.8% had stocks that could last for at least two weeks, other things remaining constant. The rest of the firms kept inventories for a month (17.6%) and over a quarter or three months (1.8%).

Figure 9.1: Distribution of Firms Based on Weeks of Inventory of Raw Materials

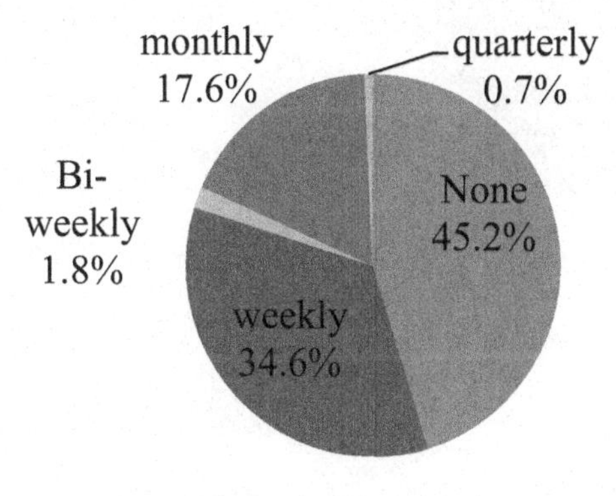

Source: Field Survey 2012

9.2.1.3 Receivables

Receivable refers to current assets (such as cash and cash equivalent) owed to a business by its clients (customers or debtors). It represents money owed by customers in exchange for goods or services that have been delivered or used, but not yet paid for. Receivables are usually due within a relatively short time period, ranging from a few days to a year. Accounts receivable is often recorded in the sales ledger and is often executed by generating an invoice and delivering it to the customer, who, in turn, must pay it within an established time frame. From the survey, all the firms that give out products on credit lines do so through 'word of mouth' and not through any formal means such as invoice or drafted agreement. In terms of duration of payment, close to 89% of the firms have payment duration less than a month. About 7.1% give out credit on net 30 days basis whiles the remaining 4.3% give out credit in the range between net 30 and net 90 days bases.

Figure 9.2: Terms of Account Receivables of Food Crop Agro-processing Firms in Ghana

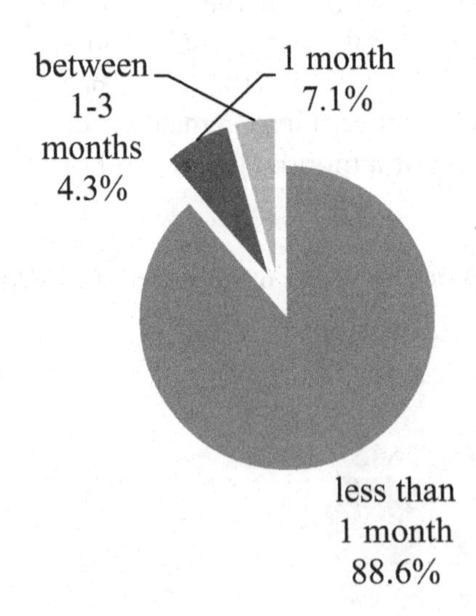

Source: Field Survey 2012

9.2.2 Fixed Assets

Fixed or capital assets refer to tangible non-current assets that have extended use, usually over one year, and that are not readily convertible into cash. Tangible assets are those holdings of individuals or companies that are real and physical in nature. Fixed assets may include land, buildings and equipment. Unlike other firms in the agribusiness value chain such as food retail firms who may not need substantial investment in fixed assets, agro-processing firms could not operate efficiently without considerable amount of investment in fixed assets. It is therefore vital for food crop agro-processing firms to look closely at their fixed assets as investments in fixed assets often involve huge financial obligations. Firms may have various fixed asset types depending on the size or scale of operation and the product types they produce. A common fixed asset held by all food crop agro-processing firms in Ghana is agro-processing complex or building. Section 5.3 provides a detailed discussion of the buildings used by the food crops agro-processing firms.

Comparing food crop agro-processing firms only on the basis of distribution of real assets such as buildings may not be adequate to show the extent of capital investment for the different firms. This is because firms may be of the same size but may be involved in the processing of different products which requires different set of capital assets. Therefore, assessment of capital investment of firms using a common basis that considers all sets of fixed assets of common firm types is more helpful. This section therefore presents the level of capital investment of food crop agro-processing firms in Ghana. Due to the paucity of data (including costs of equipment and related depreciation), only few parameters could be estimated for the purposes of comparing the extent of capital investment among food crop agro-processing firms in Ghana. The section will also discuss estimates for the aggregate value of owned and hired fixed assets, fixed capital per worker, and capital-output (k/O) ratio.

9.2.2.1 Aggregate Value of Fixed Assets Owned and Hired

The aggregate value of fixed assets owned and hired serve as a fundamental basis for comparative analysis of the level of capital investment among food crop agro-processing firms in Ghana. From Table 9.1, very small and small scale food crop agro-processing firms rank high in terms of value of assets owned by the firm. This is very surprising and contrary to the expectation that medium-scale firms should have the relatively higher average value of fixed assets. This may be due to either under estimation of the values of the assets, fear of providing accurate information that may be used for assessing corporate taxes or the higher percentage of assets that are rented by medium-scale firms. For instance, only 37.7% of the fixed assets used by medium-scale firms are owned. This is in comparison with 80.1% of the fixed assets that are owned by small scale agro-processing firms. Similarly, 79.2% of all the assets used by very small sized food crop agro-processing firms are owned by the firms. Medium-scale firms also constitute the firm categories with highest costs of assets hired or rented.

An interesting trend depicted by the results is that unlike micro sized firms who may not have the financial capacity to acquire most of the fixed assets, medium sized firms may have the capacity to acquire fixed capital. Yet they resort to hiring of equipment probably as a means to avoid maintenance cost and other cost associated with owning fixed assets. It may also be due to the tax benefits of hiring or leasing assets which are deductible as expense rather than accounting for depreciation of outright capital investment. Conversely, the results may imply huge capital required for equipment of medium sized firms for which they lack the capacity to acquire. Very small and small firms in their quest to expand, however, acquire more fixed assets.

Table 9.1: Value of Assets Owned and Hired by Food Crop Agro-processing Firms

	Average value of assets owned	Yearly cost of hiring machines/ cost of machines	Average value of assets owned and Hired	Percentage of Assets Owned	Percentage of Assets Hired
micro	3045.38	2245.75	5291.13	57.60%	42.40%
very small	5700	1495.5	7195.5	79.20%	20.80%
small	5250	1307.18	6557.18	80.10%	19.90%
medium	2500	4124.13	6624.13	37.70%	62.30%

Source: Field Survey 2012

In terms of product types, firms involved in processing of alcoholic beverages has the highest value of fixed assets they own for their operations. However, on the basis of percentage of the total assets owned, they come next to non-alcoholic beverages (94.5%). The results also show that all the assets used by vegetables, and soaps and cosmetics processing firms interviewed were hired. Firms processing vegetables have the highest value of fixed assets rented. Alcoholic beverages and roots and tubers agro-processing firms have the highest investment in fixed capital (owned and rented).

Table 9.2: Distribution of Value of Assets Owned and Hired by Food Crop Agro-processing Firms

	Value of assets owned	Yearly cost of hiring machines/ cost of machines	Value of assets owned and Hired	Percentage of Assets Owned	Percentage of Assets Hired
Roots and tubers	5732.38	2431.16	8163.53	70.2%	29.8%
Cereals	91.00	4920.75	5011.75	1.8%	98.2%
Vegetables	0.0	5060.00	5060.00	0.0%	100.0%
Oil and Oil by-products	1110.70	1866.09	2976.79	37.3%	62.7%
Soaps and cosmetics	0.0	638.80	638.80	0.0%	100.0%
Alcoholic beverages	6500.00	1875.71	8375.71	77.6%	22.4%
Non-alcoholic beverages	2000.00	115.50	2115.50	94.5%	5.5%

NB $1 = GHC2.01

Source: Field Survey 2012

9.2.2.2 Fixed Capital per Worker

Assets are operated by humans who apply their entrepreneurial skills to produce firms output. In the agro-processing industry in particular, assets are needed to optimize labor productivity, efficiency and quality of products. The proportion of assets available per worker in a firm is crucial as it determines the quality and efficiency of work in the firm. For firms to attain high capital to worker ratio implies that the firms must continuously invest in capital assets or reduce the number of employees. From Figure 9.3, capital to worker ratio declined as firm sizes increased. Contrary to normal economic practice, this may indicate that as firms grow, investment in overall capital assets decline. On the contrary, it may also show that micro-sized firms have underemployment or medium sized firms have over employment relative to capital input.

Figure 9.3: Fixed Capital per Worker

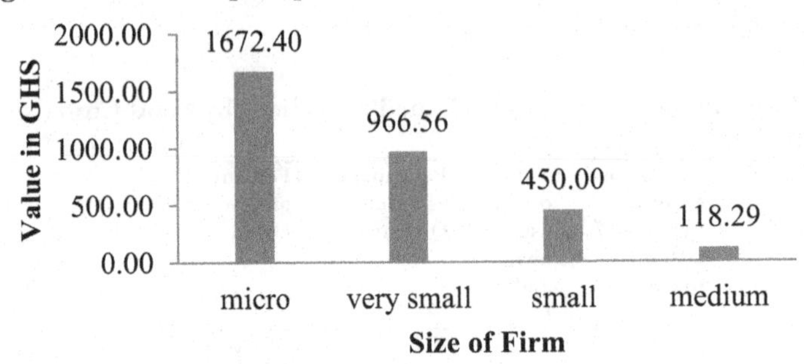

Source: Field Survey 2012

Also from figure 9.4, vegetable processors have the highest capital per worker. This is followed by cereals and alcoholic beverages. Non-alcoholic beverages as well as soaps and cosmetics processors have the lowest capital per worker value.

Figure 9.4: Distribution of Fixed Capital per Worker based on Product Processed by Firm

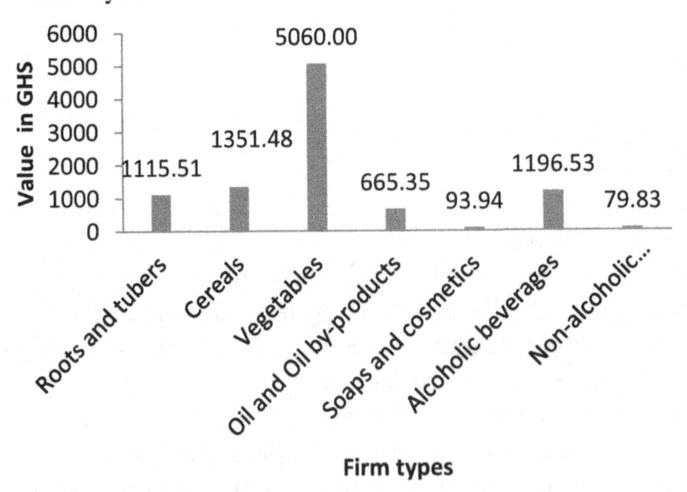

Source: Field Survey 2012

9.2.2.3 Firm Capital Output Ratio (k/O)

Capital to output ratio gives an indication of the nature and state of assets owned or used by the firms. If capital is being used efficiently, the ratio will be low which will signify that the value of capital relative to output is low or small value of capital is being used to generate a higher value of output. From the results in Table 9.3, the capital output ratio increases as the firm sizes decreases. For example, micro-scale firms used greater unit value of capital to a general comparable value of output. This may reflect the deteriorating nature of assets used by the micro, very-small, and small sized food crop agro-processing firms. Thus, medium sized firms' assets may be the best state compared to the rest of the firms or they are also able to use capital more efficiently to generate relatively greater output.

Table 9.3 Firm Capital Output Ratio (k/O)

	Capital output ratio
micro	0.2644
very small	0.1939
small	0.0451
medium	0.0042

Source: Field Survey 2012

Comparing Figure 9.4 and Table 9.4 reveals very interesting findings. Although firms processing vegetables have the highest fixed capital per worker, they generate relatively lower output with the same unit value of capital compared to firms in the other product categories. Based on the results in Table 9.4, vegetable processing firms have the least capital investment whereas alcohol producing firms have the greatest capital investment.

Table 9.4 Capital Output Ratio (k/O) of Firms based on type of Product processed

Product Categories	Capital per Output
Roots and tubers	0.0372
Cereals	0.1868
Vegetables	1.6322
Oil and Oil by-products	0.1707
Soaps and cosmetics	0.0871
Alcoholic beverages	0.0309
Non-alcoholic beverages	0.2450

Source: Field Survey 2012

9.3 Conclusion

Improved capital investment will be critical if food-crops agro-processing firms in Ghana are to improve the quality and competitiveness of their products. Most of the firms interviewed do not carry inventory due to lack of capital for the raw materials and also for the space and facilities for storing the inventory. For those who carry inventory, the study found that most of them do not have efficient inventory management systems because of poor coordination between production and marketing departments.

It was noted in section 8.9 that majority of the firms interviewed have credit sales by verbal agreements. Thus, there are no formal credit agreements. Consistent with this noted informality, this chapter notes the related poor accounts receivable management. Moreover, majority of the firms operating at micro, very small and small-scale levels have limited record of cash flows. These practices are indicators of poor financial record keeping skills and inefficient cash management systems. There will be the need for training to enhance the capacity of these firms for inventory management, cash and account receivable management.

The study found a very interesting situation, which is most medium-scale firms prefer to hire equipment for their operations rather than purchasing them. Some of the reasons behind this business decision have to do with the tax benefits of hiring or leasing assets which are deductible as expense rather than accounting for depreciation of outright capital investment. It was noted during the field survey that most of these firms hire the equipment from other firms. There are very few leasing companies providing such transactions. This provides a great opportunity for complementary business in lease finance and operation lease transactions that will support firms who may not have the adequate financial capacity to purchase the equipment for their operations.

From the study, there is increasing capital efficiency as firms expand their scale of operation. Micro-, very small-scale firms have poor capital efficiency. This may be due to the deteriorating and rudimentary nature of assets used by most of the firms operating at these scales. These firms will need improved production assets. There is also the need to enhance their capacity to use capital more efficiently to generate relatively greater output.

CHAPTER TEN: FINANCIAL RATIOS ANALYSIS OF FOOD CROPS AGRO-PROCESSING FIRMS

10.1. Introduction

Financial Ratio analysis is a process of determining and interpreting relationships between the items of financial statements to provide a meaningful understanding of the performance and financial position of an enterprise. Ratio analysis is an accounting tool to present accounting variables in a simple, concise, intelligible and understandable form (Vos, 1997).

Financial ratios are very essential to business owners. They provide information that help business owners to make crucial decisions for the business. Financial ratios can be very valuable tool for business decisions. For example, they could assist in deciding whether or not to purchase a new piece of equipment, take on a new loan or reduce rising operating expenses. A business operating based on decisions from solid and credible financial ratios will not suddenly run out of cash as illustrated in Quote 10.1. A firm should be able to forecast its cash flow accurately, manage its working capital, accounts receivables and payables if it consistently analyzes its financial ratios.

Financial ratios are very useful in understanding financial statements. Such ratios are useful in simplifying accounting figures. A long array of accounting figures can be made understandable by summarizing the figures using accounting ratios. Management can know the profitability,

Quote 10.1

SOUTH TONGU: Aveyime rice project faces imminent collapse

Rice farm

Management of Prairie Volta Rice at Aveyime in the Volta region says the company will grind to a halt in two months, if government does not support it.

Joy News has learnt the company has harvested its last rice yield and once milling of that is done, there will be no more paddy rice to mill. This is because the company in which government owns 30 per cent shares, has run out of cash and is unable to plant new seeds. Source: Myjoyonline. 2013 financial position and operating efficiency of an enterprise. These ratios measure the financial health of the enterprise and assist management to access the financial requirement and the capabilities of the business. Financial ratios are used to compare the risk and return of different firms in order to help equity investors and creditors make intelligent investment and credit decisions (Drake, 2001).

Another important benefit of financial ratio is its use in business planning and forecasting. The course of action in the immediate future is decided based on the trend of the ratios. There are times when though the overall business performance may be quite good, there may be some weak spots in some units of a business. Financial ratio can help locate those weak spots and provide appropriate remedial action. For example if the firm finds that increase in distribution expense is more proportionate to the results achieved in sales, these can be examined in detail and appropriate decisions could be executed to remove any wastage and to cut distribution costs.

This chapter estimates financial ratios to evaluate and assess the profitability and financial position among agro-processing firms in Ghana. The discussion of the financial ratios will be useful for comparisons between firms operating at different scales, types of major products and regional locations. The approach will focus on comparing the average financial ratios of firms in different categories rather than comparing the performance of a particular firm with other firms in the industry.

10.2. Types of Financial Ratios

There are several types of financial ratios that have different use and interpretation. They include liquidity, activity, financial structure, coverage, solvency, profitability and proprietary ratios.

It is important to note that not all financial ratios are significant to all businesses. The analysis undertaken will depend on the needs of the user. If a supplier wants to know if payment will be received on time for the delivery, emphasis will be placed on the liquidity ratios of the specific customer. A general manager who wants to keep track of operating expenses, cost of goods, and other operational details of the company may consider using activity ratios. One of the most difficult issues in financial analysis is focusing on the information which has meaning for a specific use without becoming lost in unrelated and inappropriate data and ratios (Vos, 1997). This chapter only focuses on liquidity and profitability ratios due to the availability of financial data of the agro-processing firms in Ghana.

Liquidity Ratio (Short Term Solvency)

Liquidity ratio measures a company's ability to pay its bills as at when they aredue, without disrupting the operations of the company. Many businesses may fail as a result of lack of liquidity. Common liquidity measures include the current and working capital ratios. Current Ratio usually measures the relationship between current assets and current liabilities. Current Assets are the assets that are either in the form of cash or cash equivalents or can be converted into cash or cash equivalents in a short time (for example, within a year's time) and Current Liabilities are repayable in a short time. Current ratio is therefore calculated as follows:

$$Current\ Ratio = \frac{Current\ Asset}{Current\ Liability}$$

In this study, the current assets for agro processing firms are sales, and raw materials whilst current liabilities are salaries paid to workers, cost of energy, yearly tax and annual loan repayment obligations.

Therefore Current Ratio for agro processing firms in Ghana is given as:

$$\frac{Sales + Inventory}{Salaries + Cost\ of\ energy + Yearly\ Tax + Loan\ payable}$$

Current ratio has a significance in showing the number of times the current assets can be converted into cash to meet current liabilities. As a normal rule, current assets should be twice the current liabilities. Low ratio indicates inadequacy of the enterprise to meet its current liabilities and inadequate working Capital whereas high Ratio is usually an indication of inefficient utilization of funds.

Profitability Ratio

Profitability ratios measure the relationships between revenues and expenses. Although the ability to generate a positive cash flow is critical for the short-term sustainability of a company, the long term financial success of a business depends on its profitability. Some frequently used profitability ratios include the rate of return ratios (based on either assets or equity), operating (gross) profit margin, net profit margin, and net income ratios. The operating profit margin is arguably one of the most important ratios for any business as it measures the profit per unit sold (Ginder & Artz, 2001). This study estimates the Return on Assets (ROA) and the Gross Profit Margin (GPM). This study uses operating profit rather than income. They are calculated as follows:

$$ROA = \frac{Net\ Profit}{Average\ Total\ Assets}$$

$$= \frac{Gross\ Yearly\ Profit - Yearly\ Tax}{Sales + Inventory + Fixed\ Assets}$$

$$GPM = \frac{Gross\ Profit}{Sales}$$

ROA may be interpreted as the amount of profit earned for assets that is employed by the food crops agro-processing firms, including the effects of borrowing. GPM is the amount remaining after paying for the cost of goods sold.

10.3. Estimates of Financial Ratios for Agro processing firms In Ghana

Figure 10.1 presents the average current ratios, return on assets, and gross profit margin of the various agro processing firms according to the scale of operations of the food crops agro-processing firms. From the figure, average current ratio has a reverse relationship to scale of operation. It increases from medium scale firms, to small, very small and then micro firms. This may be due to the relatively low current liabilities of firms proportional to current assets at lower scale of operations. As firms expand, their current liabilities increase faster than current assets. The average current ratio for micro, and medium scale firms are 7.39 and 2.08 respectively.

However, the appropriate current ratio is specific to the particular industry. It is a general acceptable business practice for a competitive firm to have a current ratio between1.5 to 3. Current ratio below 1.5 may indicate the challenges that the firm will have to meet in terms of current liabilities. Current ratio above 3 may signify that the firm is not efficient in managing its current assets or is not taking advantage of appropriate short-term leverage strategies. From the findings, medium scale firms have current ratios within the generally acceptable range. The high current ratio of micro-scale firms may indicate their inefficient use of available credit.

The return on assets follows almost the same trend as the current ratio. It is seen from Figure 10.1 that the average return on asset for micro-scale food crops agro-processing firms is higher (1.19) than small agro processing firms (0.46) which is higher than very small scaled firms (0.24) and the medium scaled firms (0.05). This means that on every 1 Ghanaian cedi (0.54 US cents) of asset, micro firms make a profit of 1.19 Ghanaian Cedis whilst Medium firm scale make a profit of 0.05 Cedis. This may be due to the relatively huge amount of initial investment required for medium-scale firms to operate efficiently. The expected value of gross margin profit for Micro, Small and Very small firms are 2.14, 0.65 and 0.47 respectively. This shows that food crops agro-processing firms operating at lower scales retain relatively higher proportion of the sales revenue after taking care of the costs of production and distribution.

Figure 10.1: Average Financial ratios of Agro-processing firms in Ghana according to their firm scale

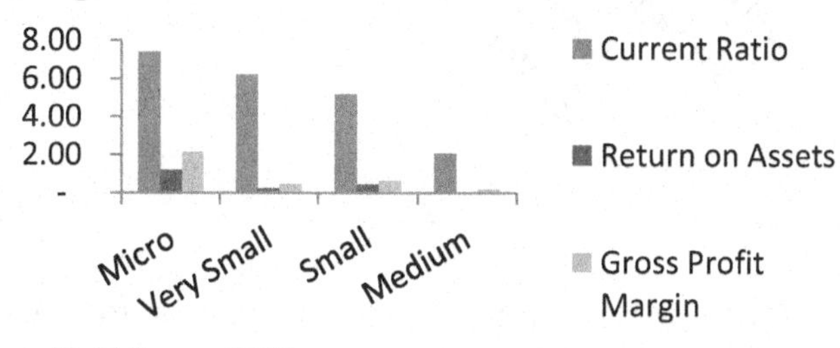

Source: Field Survey 2012.

Figure 10.2 presents the average current ratios, return on assets, and gross profit margin of the various agro processing firms according to the crop type. The average current ratio for firms processing cereals is above 10.0. This may indicate very inefficient use of current liabilities or poor management of current assets. Who goes into business to just keep current assets such as cash or inventory. Firms processing vegetables have current ratios within the generally accepted range but also have negative profit margins. This may be due to the low capital efficiency and the high costs of processing vegetable products discussed in section 9.2.2.3. It may also be due to the relatively low markup of processed vegetables. Comparing Figure 8.6 with Figure 10.2 reveals very interesting scenario. Although firms processing alcoholic beverages have the relatively higher annual sales, greater percentage of the sales revenue goes to costs of goods sold and other sales expenses. Both average gross profit margin and return on assets are lower than firms in the other categories with the exception of those processing vegetables. Higher sales revenue does not translate into higher gross profit margin for these firms.

Figure 10.2: Average Financial ratios of Agro-processing firms in Ghana according to their Crop Produced

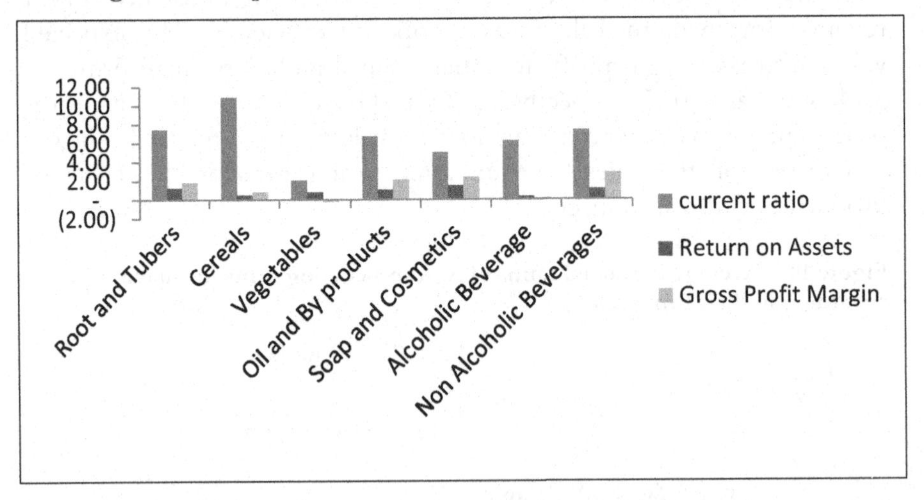

Source: Field Survey 2012.

The current ratios, return on assets, gross profit margin of food crops agro-processing firms in different regions are presented in Figure 10.3. Agro processing firms in Northern region have the highest current ratio of 9.92, followed by firms in Brong Ahafo region (7.68) and Central Region (7.65). Agro processors in Eastern Region have the lowest current ratio indicating the possible high current liabilities of firms in this region. Food crops agro-processing firms in Western Region have the highest average profit margin and return on assets.

Figure 10.3: Average Financial ratios of Agro-processing firms in Ghana according to their Location

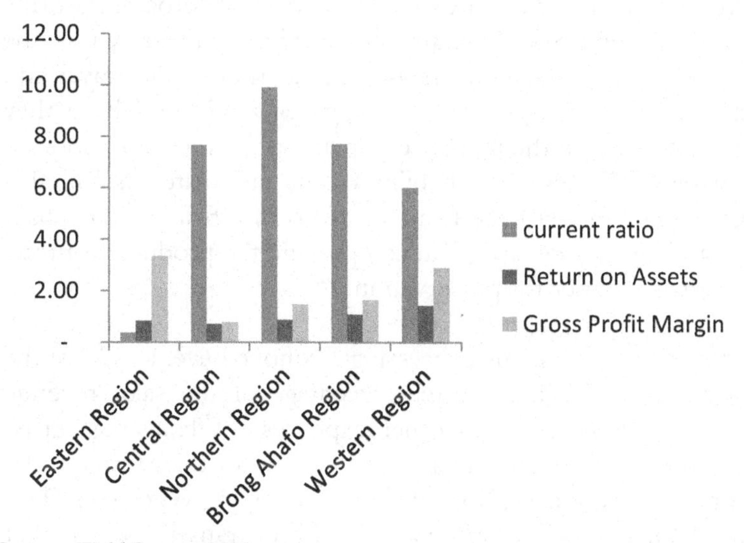

Source: Field Survey 2012.

10.4. Conclusion

Successful firms develop the capacity to persistently monitor the performance and risk of their business operations. Financial ratios play significant role in such performance monitoring. This chapter has discussed some major financial ratios for assessing the liquidity and profitability of the food crops agro-processing firms.

The study notes an unreasonably high current ratio for most of the firms operating at lower scales. For example, micro-scale firms have current ratio above 7. Consistent with normal business practice, current ratio above 3 may signify that the firm is not efficient in managing its current assets or is not taking advantage of appropriate short-term leverage strategies. Medium scale firms have current ratios within the generally acceptable range. The high current ratio of micro-scale firms indicates the need to improve the capacity of these firms for efficient use of available credit.

The study found that firms processing vegetables have negative profit margins. This may be due to the low capital efficiency and the high costs of processing vegetable products noted in section 9.2.2.3. It may also be due to the relatively low mark-up of processed vegetables. This may partly explain the reason for closure of several vegetable processing firms such as the Northern Star Tomato factory in Pwalugu. Until the government and relevant stakeholders implement strategic incentives and competiveness schemes for firms processing vegetables in Ghana they will not be able to survive the intense competition from cheap imported canned vegetables. The high perishability of the raw materials will also require governments to assist the firms to invest in efficient value chain management and storage facilities that can procure the produce from the farmers in a timely manner for preservation.

It was noted that although firms processing alcoholic beverages have the relatively higher annual sales, greater percentage of the sales revenue goes to costs of goods sold and other expenses. Both average gross profit margin and return on assets are relatively lower than firms in the other categories with the exception of those processing vegetables. This indicates that higher sales revenue may not always translate into higher gross profit margin for these firms. Generally, most of the firms have low profitability ratios. Some of the firms are just on the threshold of collapse. Because most of them do not have business plans and do not monitor the performance and risk of the business, they just operate on a subsistence basis. The government and financial institutions should not only focus on providing credit to these firms but should also enhance their capacity for business planning, operational and financial management.

CHAPTER ELEVEN: ANALYSIS OF SUPPLY OF FINANCIAL PRODUCTS AND SERVICES TO FOOD CROPS AGRO-PROCESSING FIRMS

11.1. Overview of Credits to the Agricultural sector:

Many studies have stressed on the importance of access to financial products and services for business development and growth. Particularly, for agricultural activities for which access to credit is critical for the rural population to move from subsistence farming to improved agricultural production, value addition and a diversified rural economy (Cui, 2013; Mahmood et al 2009; Koenig & Doye 1999). Cui (2013) notes agricultural credit to be an integral part of the process of modernization of agriculture and commercialization of the rural economy.

As indicated in Chapter seven (7), most food crops agro processing firms have low productivity with a general decreasing return to scale. Increasing productivity requires that firm owners implement modern and innovative technologies in the production system. However, most of these agro processing technologies require high capital investment which often cannot be easily provided by firm owners themselves. Financial resources from friends and families to supplement owners' equity are usually not adequate for such capital investments. The income that some food crops agro processing firm owners obtain is sometimes not sufficient to even cover their production costs.

As a result, most agro-processing firms produce under small scale due to financial constraints. In order to improve this situation, the government and some private financial institutions are making the efforts to provide credit to these firm owners to help facilitate purchases of adequate inputs and required technology and to improve production. It is hoped that these credit schemes may provide them the opportunity to increase their profit margins and to improve their standard of living.

This chapter will present some of the credit schemes that may be relevant to food crops agro-processing firms to receive financial resources for their business operations. It will first discuss relevant credit schemes established by the government such as the Export Development and Agricultural Investment fund (EDAIF), the Venture Capital Trust Funds (VCTF), Microfinance and Small Loans Centre

(MASLOC) and other financial products and services including the Ghana Agricultural Insurance Program (GAIP). It will then discuss financial products and services from private financial institutions such as banks (including rural banks), microfinance institutions, insurance and leasing companies.

11.2. Export Development and Agricultural Investment Funds (EDIF/EDAIF)

The Export Development and Investment Fund (EDIF) was established by Act 582 in October 2000. EDIF's mandate was to provide financial resources for the development and promotion of the export trade of Ghana. However, it was amended in October, 2011 under Act 823 of Export Development and Agricultural Investment Fund (EDIAF). Following its amendment, EDIF, now EDIAF's mandate encompasses provision of financial resources for the development and promotion of agro-processing in the country (EDIF, 2009).

The role of EDIF was to complement the efforts of public institutions to support the export trade by the private sectors. EDIF had two main accounts (a credit facility account and the export development and promotion account) out of which over 50 percent was used to support some agricultural activities like the cultivation and processing of mangoes, pineapples and shea nuts (EDIF, 2009; EDIF, 2011). Realizing the importance of agriculture in the Ghanaian economy, the agricultural component of the EDIF has been increased in the EDAIF. These activities include provision of credit guarantee through designated financial institutions to persons in agriculture and agro-processing sectors and appraisals and studies necessary to determine areas of the agriculture and agro-processing sectors that need intervention (EDIF, 2011). The agriculture and agro-processing related aspect of the EDAIF is divided into two components for ease of operational purposes. These are the Agriculture and Agro-processing Development Facility (Grant facility) and the Agriculture and Agro-processing Credit Facility.

These facilities seek to support activities in the development and promotion of agriculture and agro-processing products and offer loans and credit guarantees through designated financial institutions. Some activities financed under the Grant facility include capacity building and research, development of agricultural infrastructure and common user facilities. It also provides support for organizing agriculture and agri-business trade oriented activities of both public and private institutions (EDIF, 2011).

Applicants eligible for the agriculture and agro-processing development facility include public and private sector stakeholders in the agricultural or agro-processing sector and limited liability companies who require funds to undertake projects that would enhance the beneficiaries of agriculture or agro-processing activities. Cooperatives and associations who can receive endorsement from MOFA can also apply for these facilities. The eligible applicants for the agriculture and agro-processing credit facility are farmers of agricultural produce relating to agro-processing, marketers of agriculture produce relating to agro-processing, marketers of locally processed agricultural produce for the domestic market and investors undertaking infrastructural projects to provide services for agriculture and agro-processing entities.

Box 11.1 and 11.2 provides some information about the export promotion and development facility (EDPF) and the credit facility (CF) of EDAIF respectively.

Box 11.1

The Export Development and Promotion Facility (EDPF) supports activities of groups and institutions in the development and promotion of export products and provision of services to the export sector.

Scope of Operation
Activities financed under the EDPF include:

- Product Development and Promotion;
- Capacity Building, Market Research and Development of Infrastructure;
- Development and Promotion of other entrepreneurial activities;
- Export trade-oriented activities of institutions.

Eligible Applicants

There are two main categories of beneficiaries of the EDP Facilities. These are:

- Public and private sector stakeholders in the export trade sector such as the Ministry of Trade and Industry (MOTI), Private Enterprises Foundation (PEF), Ghana Export Promotion Council (GEPC), Federation of Association of Ghanaian Exporters (FAGE), Council for Scientific and Industrial Research (CSIR), etc.
- Association of Cooperatives/ Associations of farmers and artisans.

The assistance provided under the EDPF is expected to benefit a group, sector or industry and not one exporter.

Submission, Evaluation and Approval of Application
Application forms are obtainable from the following sources:

- Download from the EDIF website, www.edifgh.org
- Pick up from the front desk of the EDIF Secretariat in Accra.
- Completed application forms, together with business plan, must be submitted to the Fund directly.
- Applications received are processed (appraised) by the Operations Department and then submitted to Management for review. Where necessary, prospective projects are inspected prior to processing for Management's review.
- Management's decision is then submitted to the Board Committee on Export Development and Promotion for consideration and subsequently to the Board for approval or otherwise.
- The Board's decision is communicated to the applicant in writing within twenty-four hours of approval.

Disbursement
For applications approved, the following disbursement procedure is followed:

- An applicant whose application has been approved and officially written to is required to respond in writing indicating his/her acceptance of the offer and any attached conditions.
- Disbursement for approved projects is made in tranches, consistent with an agreed implementation schedule and progress of work.
- As much as possible, the Fund does not disburse funds directly to beneficiaries (especially with respect to Associations /Cooperative of farmers and artisans). Beneficiaries are expected to submit invoices covering items they wish to purchase to the Fund for approval and for direct payment to be made to suppliers. This is to ensure that facilities are not misapplied by beneficiaries.

The Fund reserves the right to revoke any approval made and all balances on such an account cancelled / frozen if at any time during the disbursement process, it is found out that the facility is being misapplied.

Monitoring and Reporting
All funded projects are subject to quarterly review by officers of the Fund from the Operations and Monitoring & Evaluation Departments. Beneficiaries are required to submit quarterly reports on the project to the Fund.

Box 11.2

The Credit Facility (loans) can be accessed through Designated Financial Institutions (DFIs). Designated Financial Institutions (DFIs) are institutions which on application are appointed by the EDIF Board to participate in the EDIF Scheme. Such institutions are incorporated under the laws of Ghana and recognized by the Bank of Ghana as carrying on the business of banking or providing credit to exporters. The DFIs include three non-bank financial institutions, namely Export Finance Company Ltd, Emprises Ghana Foundation and Exim Guaranty Company Limited.

Unlike the other DFIs who on-lend the Fund's facility to clients, Exim Guaranty provides credit guarantee services to some beneficiaries of the Fund's credit Facility.

Eligible Applicants
Individuals, corporate exporters and producers of export goods are eligible to access the Credit Facility for loans are:

- Export Marketers;
- Producers who are also Exporters;
- Producers of export goods (manufactured items/ agricultural produce); and
- Investors undertaking infrastructural projects to provide services to exporters.

To benefit from the Fund, companies or enterprises must be wholly Ghanaian owned or partly owned but with Ghanaian majority shareholding. Both existing and new businesses qualify for EDIF assistance under this Facility.

Evaluation and Approval of the Credit Facility

An applicant seeking credit facility from EDIF must obtain a prescribed application form in triplicate from a DFI of his/her choice. One completed form with supporting documents is submitted to the DFI and one to EDIF as advance information copy and one retained by the applicant for his/her records. The required supporting documents are:

- Business plan / Feasibility study/ Cash flow statement;
- Export Order/ Export Contract;
- Audited accounts for past three years or Statement of Affairs (Sole proprietorship) prepared by competent accounting firm;
- Security: copy of title deed, valuation report or other collateral security; and
- Certificate of Company Registration and Regulations.

The DFI will evaluate, appraise and give provisional approval to an application and submit its report and recommendations to EDIF. After review, final approval or otherwise shall be granted by EDIF. For successful applications, EDIF shall lend the specified amounts to the DFIs who shall on-lend same to the applicant. The DFIs bear full credit risk for loans granted.

Interest Rate

A major attraction of the EDIF Credit Facility is the low interest rate. The current interest rate is 12.5% p.a. This is to enhance the competitiveness of the country's goods on the international market.

Repayment Period

- The Credit Facility is granted over the following periods:
- Short term – for a period not exceeding twelve months;
- Medium term – for a period not exceeding five years; and
- Long term – for a period exceeding five years.

Some of the current challenges with EDAIF include lack of awareness, under-utilization of funds and the difficulty in monitoring participating banks to ensure that they are lending to agro-processing firms at the approved rate (GNA, 2013). As the Minister for Trade and Industry noted, some banks are lending EDAIF funds at 23% (Daily Graphic, 2013). There is also the suggestion to have an equity facility within EDAIF.

11.3. Rural and Agricultural Finance Programme (RAFiP)

The RAFiP is a joint initiative by the International Fund for Agricultural Development (IFAD), the Government of Italy and the Government of Ghana. The implementing agency is the Ministry of Finance and Economic Planning. RAFiP provides improved access to financial services, technical assistance and risk management instruments to rural population- particularly, smallholder farmers and micro entrepreneurs. By connecting farmers and rural micro enterprises with rural finance institutions, RAFiP seeks to support and improve sustainable livelihoods for the most vulnerable segments of the rural population, particularly women and young people (IFAD, 2013). It strengthens agricultural value chains and provides smallholder farmers with an opportunity to maximize the productive potential of their lands. RAFiP is partnering with rural microfinance institutions (MFIs) in Ghana to enhance institutional performance, public outreach and client orientation in rural Ghana (IFAD, 2013).

11.4 Microfinance and Small Loans Centre (MASLOC)

The Microfinance and Small Loans Centre (MASLOC) was established by the Government of Ghana as the microfinance apex body responsible for implementing the Government's microfinance programs targeted at reducing poverty, creating jobs and wealth.

MASLOC has been particularly mandated to hold in trust Government of Ghana and/or Development partners' funds for the purpose of administering micro and small-scale credit programs and to provide, manage and regulate approved funds for microfinance and small scale credit, loan schemes and programs (MASLOC, 2010). It provides business advisory services, training and capacity building for small and medium scale enterprises (SMEs) and collaborates with other institutions to provide relevant SMEs with the required skills and knowledge in managing their businesses efficiently and effectively (MASLOC, 2010).

Box 11.3 provides some information on types of MASLOC loans, the eligibility requirements, terms and the target beneficiaries.

Types and Eligibility of MASLOC Loans

TYPES OF MASLOC LOANS

The Micro-credit or Group Loans

Under the micro-credit scheme, the main beneficiaries are groups/cooperative societies, each consisting of a minimum of 5 and a maximum of 25 members. An individual within a group can access a minimum of GH100 to a maximum of GH500. The group solidarity mechanism is applied in this credit scheme. This means the whole group is held liable for the repayment of the loan. Thus, until every member within the group has finished paying, the group is considered not to have paid back their loan.

The Small or Individual Loans

In the case of the small or individual loan scheme, an individual can access a minimum loan of GH, 1000 and a maximum of GH10000.Under this scheme, the loan beneficiary must provide an acceptable security, in addition to a personal guarantor who must be in a position to redeem the loan in case of default.

Wholesale Lending

With this scheme, MASLOC grants loans to Microfinance Institutions (MFIs) for on-lending to small and micro businesses.

*All MASLOC loans are for a short period not exceeding 12(twelve) months within which they have to be re-paid with interest. Thus, economic activities of long gestation periods are not supported

ELIGIBILITY CRITERIA FOR MASLOC LOANS

Micro-credit or Group Loans

- The Group or cooperative society membership must be between 5 and 25;
- Individuals in the groups must be between 18 and 65 years and of sound mind;
- The Group shall have common production or operational interest; and
- The group shall have its own leaders – mainly the Chairman the Secretary, the Treasurer and with internal rules and regulations;

Small Loans

- Application must be between 18 and 65 years and of sound mind;
- Must have an ongoing business venture or project;
- Must have sound knowledge and considerable experience in the business venture project;
- Start-up project or business must be viable and capable of generating employment.

*Because of the kinds and levels of people it deals with, MASLOC does not generally require its customers to provide any business plan before accessing the loan.

Wholesale Lending

- Must be a recognized entity registered under the laws of Ghana;
- Must be committed to poverty alleviation;
- Submit a business plan or proposal;
- Submit operational manuals especially credit;
- Provide current audited or unaudited financial statements.

HOW TO ACCESS MASLOC LOANS

To access MASLOC loans one must go through the following processes:

1. Submit a written application to any of our District Office nearest you, stating the loan amount and the purpose of the loan;
2. You will be invited by one of our loan officers for preliminary assessment;
3. You will be advised on best practices, MASLOC's interest rates and other information you need to know;
4. Should you qualify for the loan, your loan applications will be processed; and
5. Applications for group and small loans should also provide two (2) passport size photographs.

MASLOC's facilities are principally targeted at the marginalized productive poor who fall mostly within the micro, small and medium enterprises sector. The main priority target groups of the intervention are women, the physically challenged (people living disabilities), and the youth especially, as well as the productive poor in general, who are operators of all kinds of small and medium scale economics/income generating activities. (Source: MASLOC, 2010).

11.5 Venture Capital Trust Fund (VCTF)

The VCTF was established by ACT 680, in 2004 by the Government of Ghana to provide credit and equity financing to eligible venture capital finance companies. Currently, there are 5 funds that are being managed as joint venture by partners. Table 11.1 shows the current funds with the partners.

Table 11.1. Venture Capital Funds and Partners.

Fund	Partners
Gold Venture Capital Limited	Gold Coast Securities Ltd
Bedrock Venture Capital Finance Limited	National Investment Bank, SIC Insurance Company Ltd
Ebankese Fund Limited	HFC Bank, Oasis Capital Ghana Union Assurance, WDBI
Fidelity Equity Fund II	SSNIT, Fidelity Capital Partners FINNFUND, SOVEC, SIFEM OIKOCREDIT, FMO
Activity Venture Finance Company	Agric Development Bank Ghana Commercial Bank

Source: VTCF, 2013.

Funds invest in all sectors of the economy with the exception of businesses engage in imports to sell. The maximum funding limit is 15% of total capitalization of a fund and a minimum of USD25,000. The VCTF ACT 680 defines an SME as a business whose total asset base, excluding land and building, does not exceed the cedi equivalent of USD 1.0million (VCTF, 2013).

Since its establishment, the VCTF has leveraged its seed funding of GH¢22.4 million to create additional GH¢40.2 million from the private sector in a Public Private Partnership. VCTF's Funds have a total of GH¢62.6 million to be invested in the SME sector. It has trained and developed the capacity of three local indigenous Fund Management companies and has provided annual training and capacity building support to the pool of VCTF Funds' portfolio companies almost annually. It has also provided education and public awareness on private equity and venture capital's role in easing access to finance in Ghana, financed about 3,500 farmers directly each year for three years running. As part of its contributions to employment, the VCTF has allegedly created 2,500 direct jobs and 4,500 indirect jobs.

11.6 Ghana Agricultural Insurance Programme (GAIP)

The Ghana Agricultural Insurance Programme (GAIP) was launched in 2011. It has developed the first agricultural Insurance for Ghana and is currently initiating key adaptation measures to climate change. GAIP provides a risk management tool for the adverse effects of climate change. It is made up of 19 Ghanaian insurances and the SwissRe insurance. The partnership with local insurance companies and a foreign reinsurance company is to ensure a solid financial foundation (GAIP, 2012).

The aim of GAIP is to protect farmers and other players in Ghana's agricultural industry from the negative economic effects of climate change. In May 2011, GAIP sold the first weather Index Insurance for maize to three financial institutions and one research organization in the Northern part of Ghana covering more than 300 small holder farmers. As at 2012 the project focused on drought insurance for maize and soya.

The Northern Region had 136 farmers being the first recipients of claims for the GAIP's drought index insurance. Currently GAIP also supply multi-peril crop insurance for commercial farmers. Plans are still being put in place to establish the area yield index insurance (GAIP, 2012). This is a very innovative risk management scheme that will offer food crops agro-processing firms some form of hedging against incidence of crop failure and shortage of raw materials. Because of their focus on the agriculture sector and relevant expertise, this paper suggests that they should also have specialized risk management schemes to cover production equipment, property and life of food crops agro-processing firms. Currently, some of the firms are using mainstream insurance products that are not tailored to their unique needs.

11.7 International Development Organizations and Multilateral Development Banks.

There are several international development organizations (IDOs) such as DANIDA and the USAID and multilateral banks such as the World Bank and the AfDB involved in promoting agricultural development in Ghana. The Alliance for a Green Revolution in Africa (AGRA) and the UN Food and Agriculture Organization (FAO) are the major international organizations involved in research and capacity development activities to improve agricultural productivity and profitability in the country.

However, in the past, most of the activities by these institutions focused predominantly on upstream activities upstream activities and less on downstream activities such as storage, processing and marketing of agricultural products. AGRA for example is currently implementing activities to improve market access of farmers particularly in the Northern Region.

As indicated in Quote 11.1, the World Bank and the USAID are jointly implementing a US$ 145 million program to support the development of commercial agriculture in Ghana (Daily Graphic, 2013). Among the activities for this program are the provision of irrigation

Quote 11.1

AGRICULTURE: World Bank, USAID Support Agriculture

Maize farm

The World Bank and the United States Agency for International Development (USAID) are providing a credit of $145 million to support commercial agriculture in the country.

The World Bank is providing $100 million, while the USAID is committing $45 million, for the project.

Source: Daily Graphic, 2013

facilities, rehabilitation of roads and other infrastructure that support agriculture, construction of warehouses, agro-processing plants and storage facilities (Daily Graphic, 2013).

One source of financial support that may be relevant to food crops agro-processing firms is the Business Sector Advocacy Challenge Fund (BUSAC).The BUSAC Fund is a Private Sector support mechanism created and funded by DANIDA, with further support from USAID, DFID (Phase I) and the European Union (Phase II). The project is aimed at contributing to the creation of a more enabling business environment for the development and growth of the Ghanaian private sector.

The BUSAC Fund was originally launched by DANIDA in 2004 as part of the broader Business Sector Program Support. The first phase of the BUSAC Fund, which was supported by DANIDA, DFID and USAID, ended in February 2010 after six years of operations. During this phase, the BUSAC Fund provided three hundred and sixty-two grants to business groups and associations spread over all the ten regions of Ghana to undertake advocacy activities aimed at improving the Ghanaian business environment.

The second phase of the BUSAC Fund started on the 1st of March, 2010. It was designed drawing lessons from the BUSAC Fund Phase I. The BUSAC Fund Phase II is being supported by USAID, European Union with DANIDA as the lead donor. The Fund is accessible through a competitive demand-driven mechanism and transparent selection of the advocacy actions proposed by Private Sector Organisations (PSOs). For BUSAC Fund Phase II, PSOs include business groups and associations and farm-based organisations (FBOs). BUSAC Fund finances, through grants, up to 90% of the cost of the Advocacy Actions that are selected in each "Call for Concept Notes." The selected Actions are then implemented by the Grantee PSOs themselves (with the help of the Service Providers they may have chosen to complement their own). Between 2004 and November 2012, the BUSAC Fund provided over 600 grants to various PSOs across the ten regions of Ghana to help strengthen their advocacy capacity. Observation and discussions during the field survey indicate that none of the food crops agro-processing firms interviewed have benefited from BUSAC fund. None of the firms was even aware of the entity. There will be the need for awareness creation for these firms to also benefit from the BUSAC Fund.

11.8 Private Financial Institutions (PFI)

Financial products and services by private financial institutions (PFIs) such as banks, insurance and leasing companies are pivotal to the development and growth of food crops agro-processing firms. The study interviewed 26 banks and microfinance institutions and had informal discussions with representatives from leasing companies, insurance companies and private equity firms. The interview with 11 major banks took place in their headquarters in Accra because of corporate policy to coordinate information about their operations and portfolios from their headquarters. The study interviewed 13 rural banks and 2 microfinance institutions in the field. Information on the percentage of credit portfolio to agro-processing firms was lacking. None of the banks and microfinance institutions could even provide the percentage of their credit portfolio to firms in the agribusiness industry. However, it was noted that most food crops agro-processing firms are not preferred customers compared to cocoa farmers and food sellers.

It was observed during the field survey that the Agricultural Development Bank Limited (ADB) and the Ghana Commercial Bank (GCB) are the predominant banks especially at the district capitals. Most of the rural banks are also located at the district capitals. Accessibility to banks for firms located outside of the district capital is a huge problem since some firm owners have to travel several kilometers for banking transactions. The risks of traveling with money over a long distanced for deposit constrain some of the banks access to formal banking services. Some of the firms, especially the micro-scale firms save with the local 'susu' agents who are informal savings and loans dealers.

11.8.1 Interest Rates by Banks offered by PFIs

The rate banks charge food crops agro-processing firms for loans have several implications to the development and growth of the firms. It reflects the costs of the capital obtained by the firm for their business activities. It is also the rate the firm usually uses to discount future cash flows from the firms activities. A very high interest rate implies high costs of capital. High interest rate also affects the financial viability of business activities financed with the credit because it lowers the present value of future cash flows.

Figure 11.1 shows the lending interest rate in Ghana compared to the rate in Nigeria, Liberia and Sierra Leone. The average lending interest rate from 2002 to 2010 in Ghana is about two times that of Liberia and Nigeria. How could firms in Ghana compete in the West African market with this outrageously high lending interest rate? The implicit high costs of capital affect the costs of production and consequently the competitiveness of food crops agro-processing firms. The wide interest rate spread indicates the low deposit rate offered by the banks – a significant disincentive to savings. The high interest rate and the wide interest rate spread points to the relative dysfunctional and prohibitive system of financial intermediation in Ghana. This should attract the attention of regulators such as the Bank of Ghana.

Figure 11.1. Interest Rate Spread (%) and Lending Interest Rate (%) in Ghana compared to Lending Interest Rate from Select West African Countries (2002-2010)

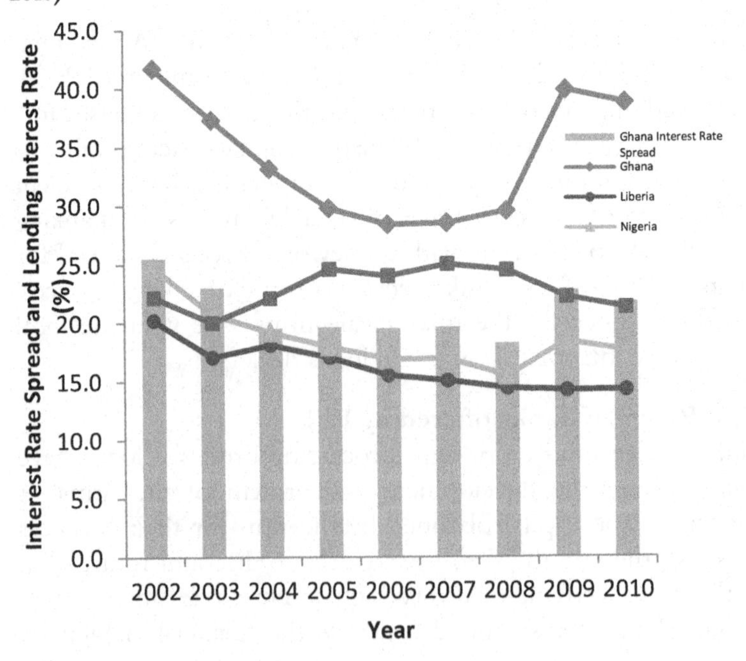

Source: World Bank (2013) and Bank of Ghana (2011)

11.8.2 Financial products and services provided by PFIs

Results from the informal consultations during the field survey indicate that the major financial product received by food crops agro-processing firms is short-term loans - below one year repayment period. The interest rates on such loans fall within the range of 4 percent to 34 percent. None of the PFIs interviewed was able to provide an approximate percentage of their loan portfolios to even general agro-processing firms. Other short-term financial products such as overdraft protection, working capital loan, bridge loan, factoring, and bankers acceptance are not offered to food crops agro-processing firms. Due to the relatively long cash conversion cycle of food crops agro-processing firms, these other short-term financial products are very critical for their operations. For example, the firm may need a working capital loan or an overdraft linked to its business account to help them manage their cash flows and to satisfy immediate cash needs for business activities. Factoring services could enable the firm to receive some financial resources from the financial institution by selling its receivables at a discount or use them as collateral for a loan to meet cash needs instead of waiting for the customers to make payments before funding its other critical business activities. About 60% of the PFIs interviewed have overdraft and working capital loan facilities but they are usually provided to cocoa farmers and businesses in construction, retail and other services. Without these short-term financial instruments, the firm owners have to resort to friends and family to meet short-term cash needs. This perpetuates the informality of their operations and financial management.

None of the PFIs interviewed have provided medium-term facilities such as lease and assets acquisition finance, real estate finance or even loans above three years repayment period to a firm processing food crops. Trade finance instruments such as buyer credit, pre/post shipment finance and letters of credit are also not provided to these firms. Some of the major reasons for the unavailability of these financial products to food crops agro-processing firms are their high risks, the high interest rates for these products and the difficulty of understanding these products by these firms -especially the micro-scale firms.

11.8.3 Credit risk analysis and loan management by PFIs

Agricultural lending usually have high transaction costs and risks because of geographic dispersion of clients, collateral problems, the small size of loans made, and the risks inherent in producing agricultural products. Most of the banks that support agro processing firms have some perceived risks when granting credits to farmers or agro-processors. The risk of giving the loans without the clients paying back is a matter of concern for several of the banks interviewed. Most of the banks interviewed did not require financial statements, business plans and articles of incorporation from the agro-processing firms before giving loans. Rather the banks require collaterals before loans are given out to them. This is the way the banks ensure that default risks are low. One PFI alleged some farmers and firm owners vacated from their location due to their inability to pay their debt. Some PFIs expressed concern about the variability of production of firms because they depend absolutely on produce from farmers that could suffer from invasion of pests, diseases and bush fires. Poor harvests due to these negative conditions affect the reliability of raw materials and subsequently their production and ability to pay loans.

Some of the banks use some recovery strategies to manage the risk of credit defaulters. They increase the collateral required and the interest rate. This has a deterrent effect of reducing the number of people who come to request for loans. Another method that is mostly used is to allow the firm owners to form groups, so that when one person defaults the group takes the responsibility for that individual. In this case the group pulls risks together and serves as mutual guarantees to minimize their individual default risks. The group deposits are used to defray default loans. Some banks call the firm owners for negotiation and follow up to the firms to help resolve defaulting cases or even take the defaulters to court as the last resort.

In order to ensure proper control and effective utilization of agricultural credit given to clients in agro processing firms, banks analyze the financial stability of the firms. They look at the firms' repayment history, profitability of the firm, liquidity of firm, management quality, client's collateral value, amount of loan and the credit term. The most important of these that were selected by the rural banks interviewed were the profitability of firms and the repayment history. This is basically because for most banks, profitability assures repayment. Banks would want to guarantee the consistency of payment of loans that have been received from firm owners before giving them more. Most of the firms are also small scaled, as noted in Chapter 4 and therefore the banks need to ascertain whether or not these firms are profitable enough to repay the debts.

Since most of the banks do not require the financial statements of firms, they send a verification team on regular visits to assess the firm's turnover amount, capital investments and management capabilities. Most banks also have records on each group and also check the trust among the groups. Some banks however look out for the sales, expenditure and cash flow of the firms before giving the clients their required loans.

11.8.4 Business financial advisory services by PFIs

Some PFIs sometimes provide business advisory services to assist customers in using the financial products such as short-term loans for the intended purpose or to generate the expected revenue. About 77% of the PFIs interviewed indicated they have staffs that are trained specifically with skills to service clients in the agricultural sector. This staffs are appointed in order to improve the financial conditions and eligibility of prospective clients in the agricultural sector. Some of the banks sometimes send staff to visit the firms monthly to observe how production systems are improving.

The officers in charge also write reports on the firms for banks to see firm's performance over a given period. Before funds are provided to clients, this trained staffs provide them with some training on financial management. The trained staff may have to explain several credit terms, conditions and processes, guiding the agricultural clients so that they do not misuse the requested funds. Some officers in charge of agricultural credits visit farmers regularly to evaluate clients' eligibility. The personnel from the bank also assist farmers to get the returns from their businesses during the harvesting seasons.

It was also noted during the interview with some of the rural banks that most clients in the agricultural sector do not use the funds for what they requested for, so the skilled staffs from the bank help them to reduce the diversion of funds and possible default. Sometimes, the funds from loan accounts are disbursed directly to the customers and vendors of the inputs and raw materials on behalf of the agro-processing firms to ensure that the loan is used for the intended purpose.

From the informal consultations with PFIs, it was noted that although some business financial advisory services are provided to particularly farmers only few food crops agro-processing firms have been provided with such services. One major reason is due to the relatively complicated production process for agro-processing firms unlike others such as farmers and food traders in the agribusiness value chain. Besides, food crops agro-processing firms will need more integrated financial and business advisory services beyond the usual assistance to fill a loan application or to secure a loan. To reduce the high informality of the agro-processing sector, PFIs will have to train their staff, for example, help these firms to execute appropriate business decisions (including capital budgeting and financial analysis), to evaluate and plan for meeting financial obligations of their business operations and to value their firms. They should also assist these firms to allocate and manage financial resources efficiently, to improve their credit profiles, to manage their liquidity (short-term working capital) and solvency (long-term financial obligations). During the field survey, it was noted that there are very few extension officers with skills and capacity to assist small businesses in business planning, business operation and management, marketing and financial management.

11.9 Conclusions

The chapter has discussed some of the credit schemes, that may be relevant for promoting food crops agro-processing in Ghana. With the exception of EDAIF that specifically targets beneficiaries that include firms in food crops agro-processing, the other schemes are more generic in nature. The ramification of the generic financial schemes such as MASLOC is the low competitiveness of food crops agro-processing firms to benefit from these schemes. Facilities offered by EDAIF could be more relevant for their assets acquisition needs, their relatively long cash conversion cycle and liquidity needs. Since the focus is also on export development, the government may have to extend the EDAIF financial products to include trade finance products such as forfaiting and letters of credit for those producing for export. It is also worthwhile to reiterate the request by the Minister of MOTI for the equity component of EDAIF. The noted lack of awareness of EDAIF is applicable to all the government-led financial schemes. There is a critical need to increase awareness and knowledge of these government financial schemes. As noted above, none of the firms interviewed have used any of these financial schemes. One of the major reasons is lack of awareness.

The major source of financial resources for food crops agro-processing firms is from PFIs such as banks and insurance companies. The high interest rate and the wide interest rate spread points to the relative dysfunctional and prohibitive system of financial intermediation in Ghana. This should attract the attention of regulators such as the Bank of Ghana. The high interest rate and the short-term payment terms is undermining the competitiveness of food crops agro-processing firms in Ghana. EDAIF should continue to monitor and reprimand (financial penalty or terminating partnership contract) offending banks that receive EDAIF funds at 12% and lend to firms at 23%.

PFIs have to provide integrated business financial advisory services to firms in food crops agro-processing. For them to effectively do this, they will have to improve their capacity and knowledge about the business, financial and investment characteristics (inventory, receivables, payables, cash flow management, operating efficiency and profitability) of these firms because they are different from farmers and food traders. PFIs will also have to understand the specific financial needs, fixed assets requirements and operational constraints of these firms. PFIs should be familiar with the decision making processes and patterns involving assets investment (fixed and variable), capacity considerations (installed and utilization) as well as their financial constraints (access, application, underwriting, disbursements, repayments). In order to reduce default rates and improve loan recovery, it will be essential for PFIs to also understand the specific business and risk profiles of firms in food crops agro-processing using for example the five major 'C's of character, conditions, capacity, capital and collateral.

MOFA may have to collaborate with PFIs and government credit facilitation programs such as EDAIF to have extension services to enhance the capacity of small businesses for critical activities such as financial records keeping, business planning, resources management (including human resources), operational efficiency and profitability planning, marketing, products and market segmentation, order sourcing, scheduling and sales management. As noted in the previous sections, most of the food crops agro-processing firms do not run their operations as business. For them to be competitive (to for example foreign companies and imported products) and to deliver quality products and services, their capacity for business planning, financial management and marketing should be enhanced. MOFA should incorporate this in their extension services.

CHAPTER TWELVE: ANALYSIS OF FINANCIAL NEEDS OF FOOD CROPS AGRO-PROCESSING FIRMS

12.1 Introduction

In Chapter nine, the study revealed the low or perhaps poor capital investment of food crops agro-processing firms in Ghana. But for food crops agro-processing firms to meet social as well as profit maximizing goals and to increase efficiency and operating capacity, firms require some form of funding. Funding sources may vary. The requirements for acquiring these funds or sometimes the lack of information on these funding sources may limit firm's ability to acquire these financial supports.

Chapter 11 has discussed some of the main formal sources of finance for food crops agro-processing firms in Ghana. These include financial institutions such as banks, micro-credit institution and private equity investors. They also include government credit schemes and interventions such as the EDIF/EDAIF, VCTF, and MASLOC. This chapter focuses on the financial needs of food crops agro-processing firms in Ghana by discussing their financial needs, access to some of these credit schemes and other financial products and services. Are food crops agro-processing firms aware of these credit schemes? Do they have access to these credit schemes? What are the major determinants of their access to credit? Do they have adequate credit for their operations? These are some of the questions that the chapter will discuss in order to improve our understanding of the supply and demand of financial products and services for food crops agro-processing firms in Ghana.

12.2 Financial Needs of Food Crops Agro-processing firms.

As shown in Figure 12.1, food crops agro-processing firms need access to credit to enable them finance assets such as buildings, processing and packaging equipment. They also need credit to help manage their working capital. To reduce the risks of loss due to fire, thefts and natural disasters, they also have to insure their fixed assets and sometimes themselves. They usually finance their fixed assets costs, production costs (including costs of inventory) and working capital (to meet other short-term recurrent expenses) from their own equity and loans from their relatives and friends. They sometimes supplement these sources of credit with high interest loans from rural "loan sharks".

Although the government has established some credit schemes such as the EDAIF and MASLOC, it is important to underscore the fact that the agribusiness value chain consists of industries with different characteristics, needs, niche and markets.

Figure 12.1 The Gap between Financial Needs and Available Funds for Food Crops Agro-processing firms in Ghana.

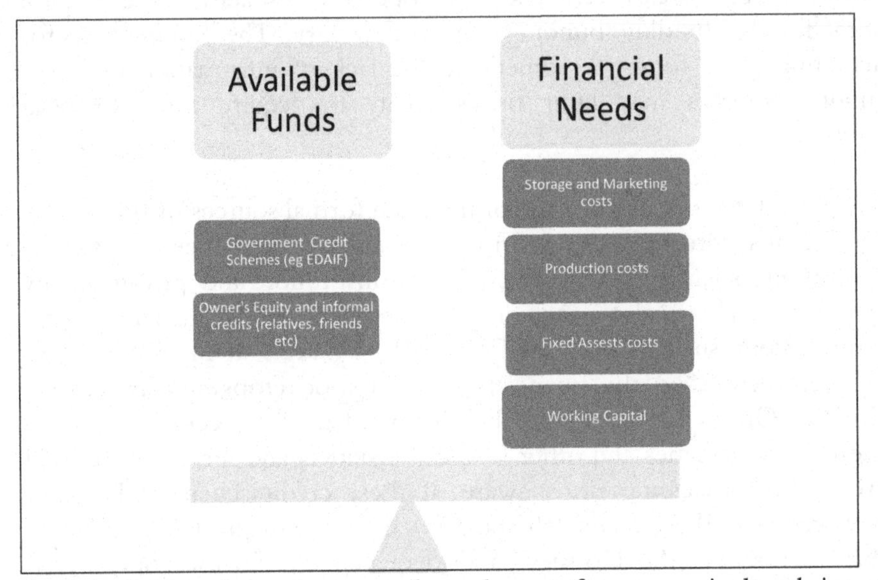

A financial product that may be relevant for an agricultural input industry may not be relevant for a food processing or a food retail industry. For example, MASLOC credits are largely short-term up to 12 months. Such a credit scheme may not be appropriate for SMEs in agro-processing who may need medium-to long-term loans to finance real assets and processing equipment with repayment periods often similar to the lifespan of the assets. Besides, unlike food traders, SMEs in agro-processing have relatively long cash conversion cycles which affect their liquidity and ability to service short term loans for operations other than working capital. Some of these financial issues are discussed in section 10.3. For food crops agro-processing firms to benefit from such public financial schemes, their specific financial needs, fixed assets requirements and operational characteristics should be taken into consideration. Moreover, financial resources from government credit scheme may not be adequate. As shown in Figure 12.1, there will still be financing gap that may require market-based financial resources from private financial institutions (PFIs) to bridge the gap.

12.3 Access to Credit

Meanwhile, food crops agro-processing firms lack access to credit and other financial services because they are often perceived as informal and high risks lenders. Their access to market-based credits from banks and other financial institutions such as factoring and leasing companies are very limited. Inadequate access to credit by agro-processing firms is one of the major constraints to the development of agro-processing industries (Derbile *et al.*, 2012, Ofei, 2004). From the field observations and discussions, among the major difficulties associated with credit acquisition by food crops agro-processing are the long bureaucratic and application processes, short moratorium and repayment duration, inadequate level of credit and high interest rate.

As shown in Figure 12.2, only 37% of food crop agro-processing firms interviewed have access to credit. Majority of the firms did not have access to credit. The results provide ample reason for the poor capital investment and growth of firms discussed in chapters 9 and 10.

Figure 12.2 Access to Credit by Food Crops Agro-processing Firms in Ghana

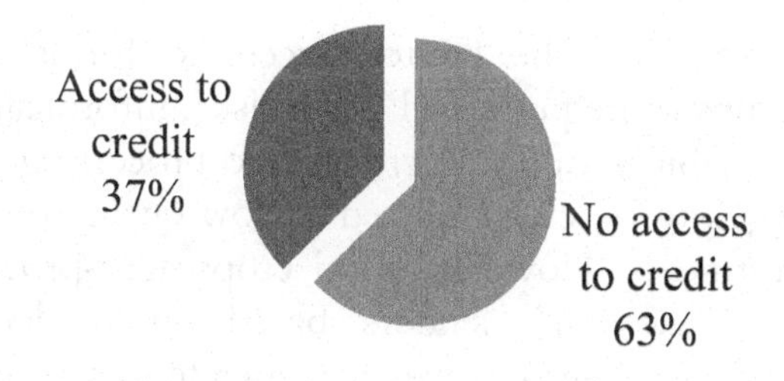

Source: Field Survey 2012

As illustrated by Figure 12.3, banks are the dominant source of credit to food crops agro-processing firms. About 47% of the food crops agro-processing firms that have access to credit obtained their credit from the banks. Also, friends of firm owners are instrumental in the provision of credit. About 23% of the food crop agro-processing firms have obtained credit from their friends. The rest of the firms accessed credit from family members (11%), government institutions (10%) and FBO's (9%). Although majority of the firms have access to credit from formal sources (banks and government institutions), the results illustrates the overall importance of informal credit sources as complements to formal credit institutions.

Figure 12.3: Sources of Credit to Food Crops Agro-processing Firms in Ghana

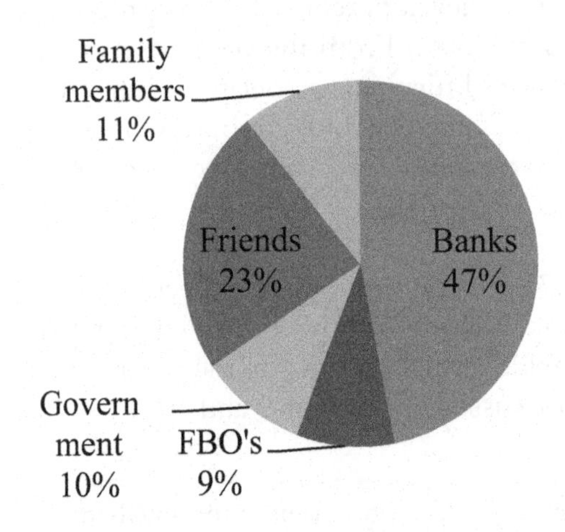

Source: Field Survey 2012

Some of the major reasons for the limited access to formal credit especially through the banks is the perceived "high risk", informality and uncertainty of business viability of food crops agro-processing firms. The industry is also alleged to be characterized by low quality products and low return on investments. However, food crops agro-processing firms should be more "low risk" lenders based on fundamental economics. SMEs in food crops agro-processing operate in a consumer defensive sector with "recession proof" demand for food and potential for steady return on investments. As shown in section 8.3, the growing urbanization and the middle class in Ghana and most West African countries has spurred demand for high quality processed food. With improve quality control, these firms will be very competitive and profitable. Access to credit will be critical for them to have the financial capacity necessary to improve the competitiveness of their products against foreign imported products.

Experiences from other countries such as Vietnam indicate that their perceived characteristics of "high risk, low return" are more cyclical and the limited access to credit constrain their efforts to improve their processing efficiency, quality standards and marketing of products (Hai, 2006). Food crops agro-processing firms could demonstrate their potential for steady liquidity, operating efficiency and profitability as well as the potential for high return on their investments if they are provided with the opportunity for improved access to credit.

12.4 Equity Financing

Raising capital is an important first step in any business operation. Usually, a company's capital funding consists of both debt (bonds) and equity (stock). Bond and equity holders earn a return on their investment in the form of interest, dividends and stock appreciation. The mandate of some institutions is to provide equity funds to firms. These firms often provide funding for specific categories of firms at a particular stage of the firm or they may finance aspects of the value chain (value chain financing). In Ghana, the Venture Capital Trust Fund (VCTF), the Business Sector Advocacy Challenge Fund (BUSAC), the Social Investment Fund (SIF) - a component of the Ghana Poverty Reduction Programme (GPRP) provides equity finance to SME's. As noted in Chapter 11, VCTF was established by ACT 680, in 2004 by the Government of Ghana to provide credit and equity financing to eligible venture capital finance companies such as the Bedrock Venture Capital Finance Limited managed by the National Investment Bank and the SIC Insurance Company Limited to support SMEs.

Figure 12.4: Sources of Firms Equity

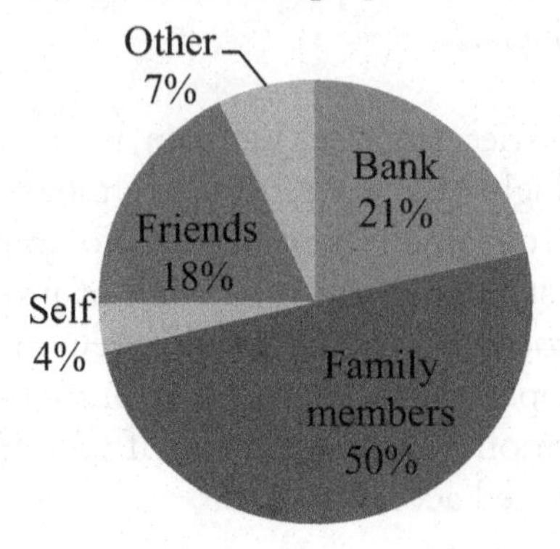

Source: Field Survey 2012

As shown in Figure 12.4, none of the firms interviewed have received equity financing from these national equity financing companies. The 21% of firms who received equity financing from banks did secured them as loans without any equity interest of the banks but the assets were used as part of the collateral. The main sources of equity funds for food crop agro-processing firms are from family members. This indicates that most firms are either not aware of equity financing through some of these financial institutions or just fail to access funding from them. There will be the need for awareness creation for these sources of financing.

12.5 Determinants of Access

This section will analyze the factors that determine access to credit by food crops agro-processing firms in Ghana. Six main variables identified in literature to determine firms' access to credit were selected and modeled using the Probit model. From the results, Wald chi square test at six degrees of freedom is significant at one percent level. This means that the coefficients of all predictor regression coefficients in the model are simultaneously not equal to zero.

The results show that education, size of firm as well as the business assets of the firm significantly determines access to credit by food crop agro-processing firms in Ghana. As shown in Table 12.1, age of the firm owner, gender and years of establishment has no significant effect on access to credit.

Table 12.1: Determinants of Access to Credit by Food Crops Agro-processing Firms

Dependent Variable: Access to credit measured as a dummy variable			
Variable	**Coefficient (β).**	**Std. Err.**	**Marginal effects (dy/dx)**
Age	-0.00916	0.006922	-0.00351
Gender	0.10491	0.237330	.0398257
Education	0.08916***	0.026639	0.03415
Firmsize	0.03161***	0.010805	0.01211
Firmage	-0.01124	0.018726	-0.00430
BusinessAssets	0.00016**	0.000066	0.00006
_cons	-0.75440	0.54545	
N	272		
Wald chi2(6)	32.93		
Prob > chi2	0.0000		
Log pseudo-likelihood	-150.0820		
Pseudo R2	0.165		

Source: Field Survey 2012

NB: ***, ** implies significant at 1% and 5% respectively

The results indicate positive significant relationship between education and access to credit by food crop agro-processing firms. Thus, other things remaining constant, a year increase in educational level of firm owners leads to 0.034 unit increases in the probability of accessing credit by food crop agro-processing firms. This is very interesting because education plays significant role in access to information and the ability to take advantage of opportunities in the formal sector where requirements such as application and provision of relevant documentation or materials form part of the procedure for credit.

Similarly, a unit increase in firm sizes increases the probability of accessing credit by 0.012 units. As discussed in section 10.3 firms at lower scale of operation have poor current assets management and are not efficient in utilizing available credit as indicated by their unreasonably high current ratios. This poor financial management is complicated by the poor record keeping that constrain the ability to submit required documentations for the credit application process. Their high degree of informality exacerbates the perceived low returns, risks and uncertainty of performance of investment in the food crops agro-processing sector.

Another significant determinant of access to credit is the assets base of the firm. From the Probit estimates, a cedi (US$0.54) increase in asset base of the firm increases the probability of accessing credit by 6.0e5 units. This suggests that as firms grow and acquire more assets, their chances of accessing credit increases. This is because they could use these assets as collateral for accessing credit.

12.6 Constraints in Accessing Credit by Food Crops Agro-processing Firms in Ghana

As shown from Figure 12.5, about 85% of the loans received by food crop agro-processing firms were short term loans. These loans have repayment duration ranging from one week to one year. Over 50% of the loans advanced to the firms had a repayment period in the range of seven to one year. Only 15% of the total credits accessed by the firms interviewed are between one to two years repayment terms. How could a firm finance the acquisition of production assets such as processing equipment that may have an operational lifespan of about 10 years? Considering the fact that short term loans do not help organizations in their long term planning, it is appropriate to have innovative and inclusive financial products and services specific to the financial needs and characteristics of food crops agro-processing firms.

Figure 12.5 Repayment Duration of Credits Assessed by Food Crop Agro-processing Firms in Ghana

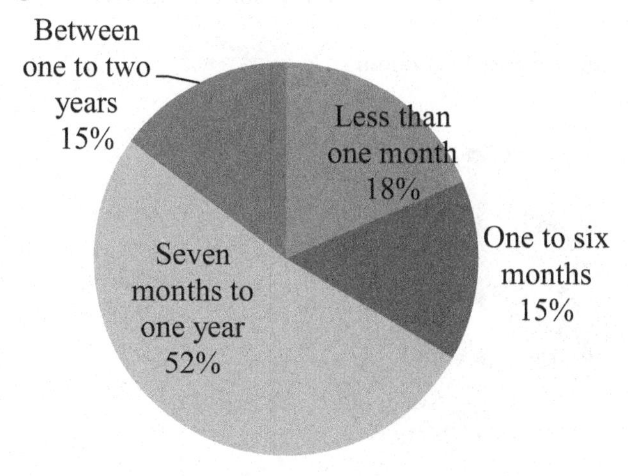

Source: Field Survey 2012

Also, interest rates remain very high. Figure 12.6 shows that about 70% of the firms secured their loans at interest rates below 10%. Considering the fact that the EDAIF credit facility has a defined interest rate of 12.5% per annum, these sources of credit offering interest rate below 10% may be more competitive than the EDAIF.

Figure 12.6: Distribution of Food Crop Agro-processing Firms Based on Interest Rates of Credits Accessed

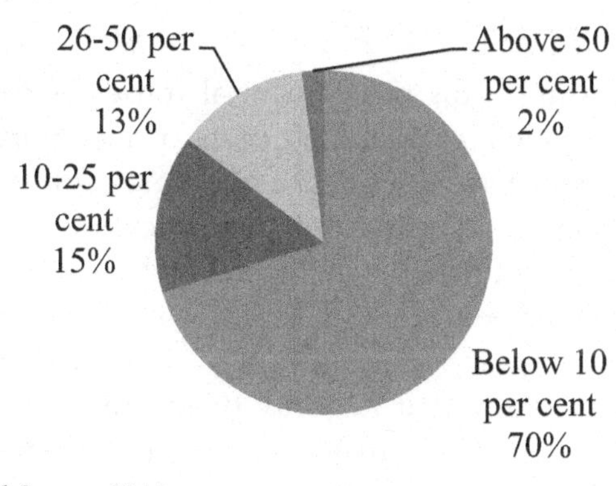

Source: Field Survey 2012

About 23% of the firms accessed credit at rates above 10% but lower than 50%. Only 2% secured loans with interest rates above 50%.

Figure 12.7: Adequacy of Credit Amount

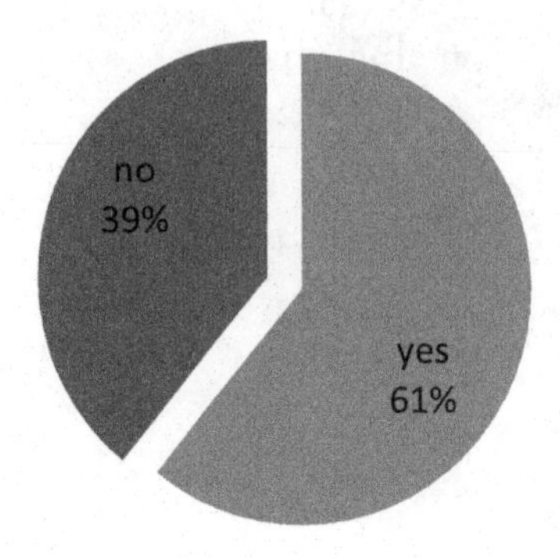

Source: Field Survey 2012

Aside the high interest rate, the amount of loans received by food crops agro-processing firms, are often inadequate. From Figure 12.7, about 40% of the firms interviewed indicated that the loans advanced to them were inadequate. Most firm owners confirmed that they were not able to use the loans to achieve the intended purpose. They either had to look elsewhere for funds to complete their project or had to misapply the funds.

12.7 Conclusions

Food crops agro-processing firms need financial resources to meet the costs of fixed assets, production (including costs of inventory) and for working capital. They also have to finance the costs for storage and marketing of their products. The chapter has discussed the financial needs of these firms, their financial characteristics and major factors determining their access to credit.

Results from the analysis show that majority of the firms do not have access to credit. This is one of the major reasons for the poor capital investment and growth of food crops agro-processing firms as discussed in the previous chapters. Although the government has established some credit schemes such as the EDAIF and MASLOC, and equity financing funds such as the VCTF, most of the firms interviewed have not benefitted from these opportunities. Some of the reasons will be the financial characteristics of these firms that make them subprime clients compared to other firms in the services and even in the agri-business value chain that have shorter cash conversion cycles and are able to service short term loans. The results from the study also indicate that credits received by these firms are inadequate, have high interest rates and very short repayment terms below one year. This underlies the generally poor capital investment, poor efficiency of conversion, poor quality of products, low competiveness and profitability of these firms discussed in the earlier chapters. Firms are not able to finance assets that will help them to improve their operations, products and profitability due to these constraints. For example, how could a firm finance acquisition of processing equipment with lifespan of about 10 years with short-term loans? Majority of the firms therefore depend on families, friends and rural informal lenders for credit.

The government of Ghana will have to partner with financial institutions to strategically plan and implement customized agricultural financing schemes targeting SMEs in agro-processing of food crops. Such schemes should be pursued to improve the efficiency of operations, the quality of financial transactions and higher debt recovery in order to reduce the high rate of defaults often associated with government-led financial schemes. Some of the avenues and sources of funding for these agricultural financial schemes could be risk capital from development financial institutions (DFIs) that has concessional terms and longer investment horizon.

The results also indicate that education, size of the firm and business assets are the significant factors determining access to credit. There will be the need for literacy and numeracy training activities that will enhance the capacity of the management of these firms to apply and provide required documentation to secure adequate credit for their operations. There are so many firms in this sector operating below appropriate scale. This makes them inefficient and less viable business ventures. The relatively high transaction cost of servicing so many micro-level firms is constraining their access to credit. They are also characterized by high degree of informality that worsens their perceived high risks, low returns and poor performance of investments. This result substantiates the recommendation in chapter 4 to consolidate particularly the micro- and very small-scale firms to make them operate at viable scale and with adequate business assets to serve as collateral for loans.

CHAPTER THIRTEEN: STRATEGIC INTERVENTION AND NATIONAL IMPLEMENTATION PLAN FOR PROMOTING FOOD CROPS AGRO-PROCESSING IN GHANA

13.1. Introduction

Food crops agro-processing firms in Ghana play significant roles in providing value addition to agricultural products and help reduce post-harvest losses that undermine productivity, profitability and food security in the nation. They also provide employment, income and alternative source of livelihood to households in particularly rural communities and help spur rural industrial and enterprise development that contribute to the diversification of rural economies from subsistence farming.

With such critical roles played by food crops agro-processing firms, the government cannot pursue a comprehensive national development and inclusive growth strategy without focusing on improving their operations, productivity, efficiency, profitability and competitiveness. In fact, this book will emphasize the opportunity available to the government of Ghana to use food crops agro-processing to rationalize the national strategy for rural industrialization, import substitution for food products, employment (including youth employment) and inclusive growth in the nation. The government should explore and commit the needed resources to take advantage of the huge opportunity available to the nation by improving food crops agro-processing. This book will provide a simple framework for intervention based on some of the major findings of the research.

13.2 Some Major Findings

In order not to make the situation too overwhelming by presenting too many challenges and constraints facing food crops agro-processing firms in Ghana, only few major findings will be highlighted in this chapter. The relevant findings are organized into three major areas – (1) Structure, Distribution and Concentration, (2) Operations, Quality, Productivity and Marketing and (3) Capital Investments, Supply and Demand for Financial Products and Services.

13.2.1 Structure, Distribution and Concentration of Food Crops Agro-processing sector in Ghana.

- Over 76 per cent of food crop agro-processing firms are owned by women.
- About 79% of the firm owners have no formal education.
- About 85 per cent of food crop agro-processing firms in Ghana are micro-sized firms. They employ about half of the workforce in the food crops agro-processing business. These micro-level firms could operate for over 10 years without expanding in scale. This is an indication of survival or subsistence production. Majority of the firms operate in simple structures with rudimentary processing technology and have low skills for operations, resource mobilization and financial management.
- Food crops agro-processing firms in the Eastern Region are highly concentrated with the few medium-scale firms controlling greater portion of the market. However, there is a moderate concentration of firms at the national level with several micro- and very-small companies that could be consolidated to make them more financially viable. The product markets with less concentration are vegetables, cereals, alcoholic beverages, soaps and cosmetics, roots and tubers.

13.2.2 Operations, Quality, Productivity and Marketing

- Most of the food crop agro-processing firms have low installed capacity which often remains underutilized during off-seasons and over-utilized during peak seasons. Among the factors contributing to low operating capacity are lack of reliable energy, raw materials, and consistent demand.
- Most of the processed products are of low quality and below required international standards. Most of the firms have not diversified their products and markets and have weak or no internal and external linkages. They also lack adequate human resource skills as well as operating capital to manage the growth and profitability of the firm.
- The productivity of the MSMEs is determined by the cost of labor, physical capital and raw materials. However the most important factor is raw materials.

- Firms are experiencing decreasing returns to scale with respect to input use due to the predominantly labor intensive and rudimentary processing technologies that generate minimum value addition.
- Productivity and efficiency of utilizing raw materials improves consistent with scale of operation. As firms expand from micro, very small to small scale, they become more diversified in their products delivery which also improves their efficiency of conversion of raw materials.

13.2.3 Capital Investments, Supply and Demand of Financial Products and Services.

- The average annual turnover of the MSMEs ranges from GHC 600 ($324) for micro-scale firms to GHC 34,560 ($18,662) for medium-scale firms. Generally, most of the firms have low profitability ratios. Some of the firms are just on the threshold of collapse.
- Most of the firms do not carry inventory due to lack of capital for the raw materials and also for the space and facilities for storing the inventory. Greater portion of the firms who carry inventory do not have efficient inventory management systems because of poor coordination between production and marketing departments.
- There is increasing capital efficiency as firms expand their scale of operation. Micro-, very small-scale firms have poor capital efficiency. This may be due to the deteriorating and rudimentary nature of assets used by most of the firms operating at these scales.
- High interest rates of loans limit the potential profitability of operations for the firms. The essential financial needs of MSMEs are medium- to long-term loans to finance real assets and processing equipment. Due to their long cash conversion cycle, low liquidity and the medium to long-term investment horizon of most of their capital investments, they will need longer repayment periods.

- Firms have low capital investments. Less than half of the firms have access to credit. Access to credit is determinant by education, firm size, and years of establishment.
- With about 85% of the firms operating at micro-scale, the subsector is characterized by high informality with lack of business planning, financial records keeping, financial analysis and monitoring of risks and performance. This high informality increases the perceived credit risks that limit firms access to credit.

13.3 The Intervention Framework.

The analysis and discussions have identified several critical issues that should be addressed in order to realize the full benefits of improving food crops agro-processing in the nation. With so many issues to deal with, how do you intervene? And what are the issues to prioritize? This book presents a simple framework to prioritize issues for intervention. The intent is to provide actionable decisions that could have immediate implementation.

The major decisions and policies are illustrated in Figure 13.1. The left side of the intervention framework presents the critical macroeconomic issue of high policy rate that is negatively affecting the competitiveness of MSMEs in food crops agro-processing. Although the analyses and discussions in this book are essentially on microeconomic issues presented at the right side of the intervention framework, there is no way that the macroeconomic issue of high policy rate could be ignored. The high policy rate does not only have implications at the national level but also at the firm level. At the national level, the high policy rate has contributed to the high yield spread that has attracted massive amount of capital into the country.

Figure 13.1. Strategic Intervention Framework for Improving Food Crops Agro-processing in Ghana.

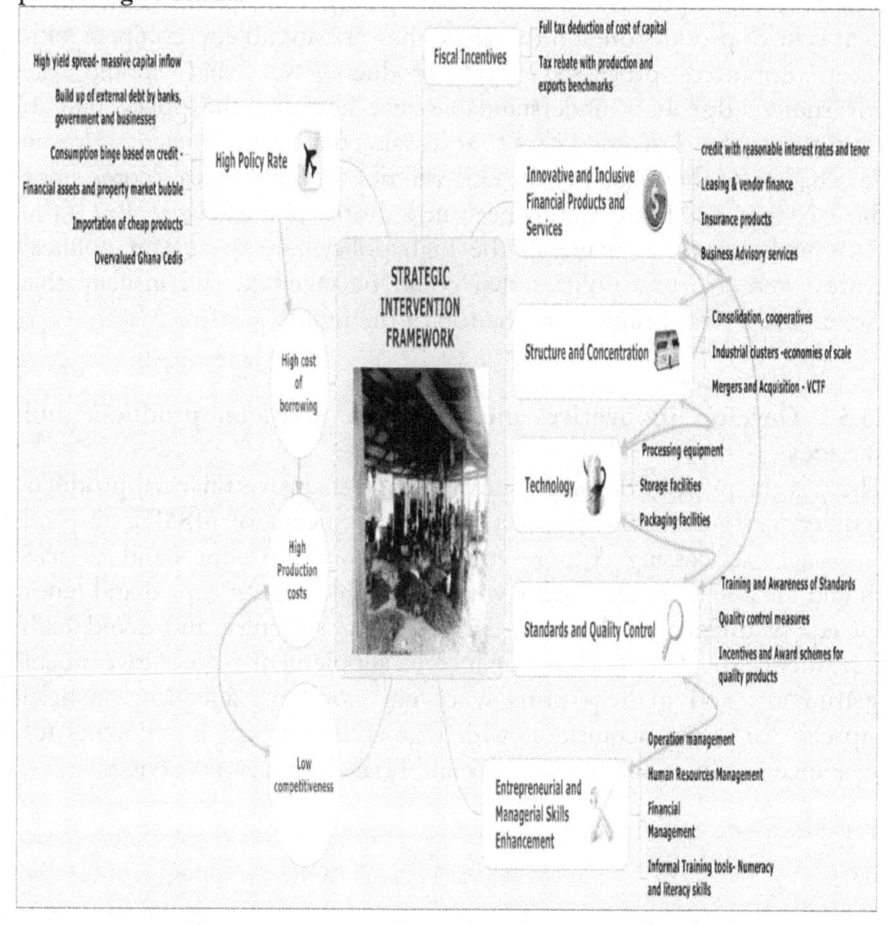

The government, businesses and banks have accumulated substantial amount of debt as a result of this capital flow. Ghana's external debt is currently about GHC 46 billion or US$22 billion (Business News, 2013). There is also massive built up of debt by businesses, banks and private individuals as a result of the consumption binge based on availability of high interest rate credit. Because the Ghana Cedi is overvalued as result of the upward pressure from the high policy rate and the high inflation of almost 13 percent (Business News, 2013) this has encouraged the importation of cheap products into the country. The overvalued currency has undermined the competitiveness of Ghanaian products. As discussed in Section 11.8 above, at the micro level, the high policy rate and the resultant high interest rate has resulted in high cost of capital for food crops agro-processing firms. The high cost of capital has increased their cost of production and therefore they are not able to compete with cheap imported processed food products. So what should the government do? It is understandable that lowering the policy rate to appropriate level could cause reversals that may include drastic devaluation of the Ghana Cedi. This will not fly for political scores since the exchange rate has currently become a thorny political issue. But if the government want to maintain the high policy rate to sustain political scores there are two policies that could be taken to still insulate this target group. This brings us to the top of the figure.

13.3.1 Develop innovative and inclusive financial products and services

There is the need to develop innovative and inclusive financial products and services customized to meet the specific needs of MSMEs in food crops agro-processing. A more integrated financial products and services should include (a) credit facility with reasonable cost of capital and tenor (b) equity finance to diversify the financing structure and avoid high dependence on debt, (c) lease finance to supplement regular investment instruments and to help firms who may not have adequate financial capacity for assets acquisition with regular investment instruments (d) insurance products and (e) business and financial advisory services.

Figure 13.2 Integrated Financial Products and Services for MSMEs in Food Crops Agro-processing.

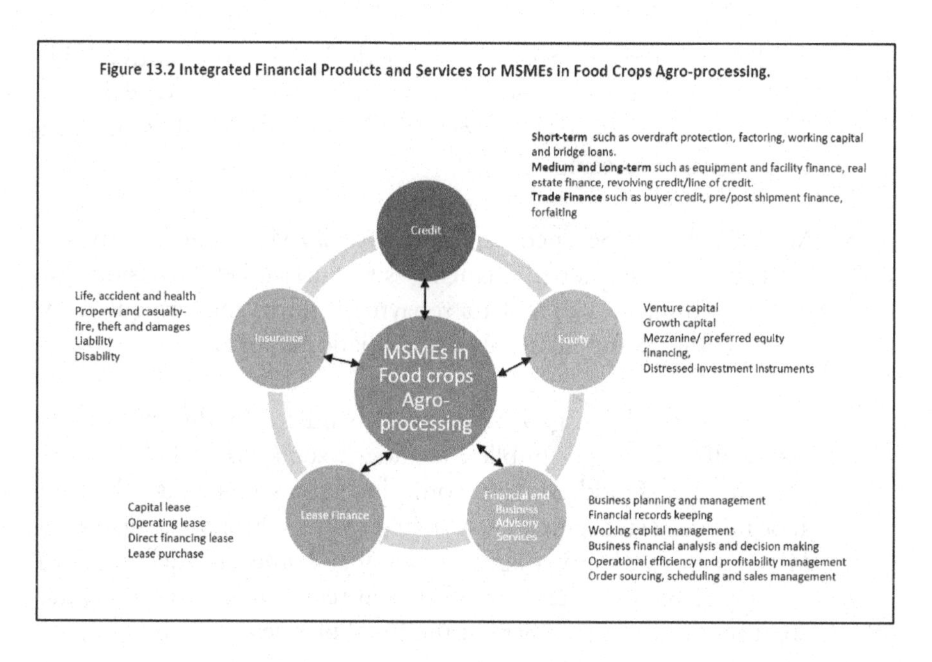

Although financial intermediation firms offer different types of loans most of them are only for short term. As noted in Section 12.6, only 15 percent of the firms interviewed have secured loans with repayment period between one to two years. None of the firms have secured loans with more than two years repayment terms. The interest rates are also prohibitive. It will be essential to have a private-public credit facility with different types of loans, different interest rates and tenor to meet specific needs such as for assets purchase or for working capital management. Currently, EDAIF's credit facility provides loans at an interest rate of 12.5 percent which is far lower than the average interest rate of about 30 percent in the mainstream financial intermediation market. However, findings from the research indicate that none of the MSMEs interviewed during the field survey have benefited from EDAIF credit facility. The credit application and approval process is also excruciatingly complicated and long.

The funds are limited and could not cater for the level of demand for loans by MSMEs in food crops agro-processing. Some recommendations to consider include:

- Enhance the EDAIF credit facility if having a new facility for agro-processing firms will not be cost-effective. Since MSMEs in food crops agro-processing are relatively less competitive with access to credit, there should be specific allocation (for example 50 percent) of the EDAIF credit portfolio to MSMEs in food crops agro-processing.

- MSMEs should be encouraged to use EDAIF credit facility for medium- to long-term financing since it makes provision for credit with more than 5 years repayment terms that are currently not available in the financial intermediation market.

- To bridge the huge gap between demand for credit and supply essentially through public facilities such as EDAIF, the government should provide some incentives to banks who will allocate at least 5 percent of their credit portfolios available to MSMEs in food crops agro-processing. Some tax incentives to encourage banks to lend to MSMEs at reasonable costs of capital are discussed below as part of the fiscal policies.

- Enhance access to equity financing. This will be essential for MSMEs that have so much debt that piling up another debt will generate high cost of capital since they will not have conventional or preferred rate and are considered 'high' risks clients. Also because of the concentration of debt in the financing structure, they may have to consider equity in order to diversify and reduce their weighted average cost of capital (WACC). Equity financing will also be necessary for MSMEs who persistently have return on the invested capital (ROIC) lower than the cost of capital or who may need some management capacity, solid business models or marketing strategy to realize residual benefits from past investment. The VCTF could help enhance the awareness and knowledge of MSMEs in food crops agro-processing for equity financing and facilitate their access to its five partner funds.

- Improve access to insurance and leasing services, including micro insurance and micro leasing financial transactions to enable particularly the predominantly micro-scale firms to handle capital investments and risk management for their operations.

- Enhance the capacity of financial institutions (FIs) and extension officers to provide financial and business advisory services to MSMEs in food crops agro-processing. FIs should be capable of providing MSMEs with integrated financial advisory services that will reduce their default rates. Currently, these services are unfortunately lacking. FIs should be trained in for example, how to help MSMEs to executive appropriate business decisions (including capital budgeting and financial analysis), how to evaluate and plan for meeting financial obligations of their business operations, how to value their firms, how to allocate and manage financial resources, how to improve their credit profiles, how to manage liquidity (short-term working capital) and solvency (long-term financial obligations). There are very few MOFA extension officers with skills and capacity to assist MSMSEs in business planning, business operation and management, marketing and financial management. As noted particularly in Chapters 7, 11 and 12, any investment facilitation program for MSMEs will not achieve significant outcomes without extension services to enhance their capacity for critical activities such as financial records keeping, business planning, resources management (including human resources), operational efficiency and profitability planning, marketing, products and market segmentation, order sourcing, scheduling and sales management. To reduce the high informality of MSMEs and to improve their competitiveness and capacity to deliver quality products and services, their capacity for business planning, financial management and marketing should be enhanced.

13.3.2. Provide Fiscal incentives for food crops agro-processing firms.

Section 10 of the Internal Revenue Acts of 2000 and its 2006 and 2012 Amendments provides for five years exemption of income tax for an agro processing business established in Ghana in or after the financial year commencing 1st January 2004. As discussed in Chapter 5 of this book, only 14 percent of the firms interviewed have been in business for 10 years or lower and could qualify for this tax exemption. For those who may not have tax exemption of their corporate income, they could have a full deduction of the interests they pay on a loan, expenses they incur for repairing equipment, buildings and facilities as well lease expenses. They could also deduct capital allowances for equipment, facilities and buildings used for the business operation. However, the high costs of raw materials, labor and other variable inputs caused by the high policy rate environment and high cost of doing business are not tax deductible. Moreover, there should be some tax incentives for FIs who will partner with the government to bridge the huge financing gap. These recommendations should be considered:

- Provide tax incentive such as tax rebates to firms that put in place measures to improve quality standards and achieve specified production and efficiency benchmarks. These tax rebates will be of great benefit to particularly those companies established more than 5 years and could not have tax exemption for corporate incomes. These tax rebates will also help improve the profitability of these companies and provide them with the financial capacity to invest in improved technology for their operations.

- Give FIs who allocate at least 5 percent of their credit portfolios for MSMEs in food crops agro-processing exemption from capital gains tax on the credit they advance to the MSMEs. This capital gains tax exemption will provide the incentive for FIs to offset the reduction in interest rates for loans advanced to MSMEs in food crops agro-processing.

13.3.3. Consolidate the MSMEs to improve economies of scale and financial viability.

As noted in Chapters 5 and 7, the predominance of micro-scale firms and their employment potential show their significance for improved household income, inclusive growth and diversification of rural economies in Ghana. However, majority of the micro-scale firms are operating below recommended capacity with indigenous and inefficient technologies. They also have weak internal and external linkages, low survival rates and characterized by high informality, lack of transitional planning and business continuity strategies to improve their scalability and sustainability. Because these micro-scale firms are scattered, there will be high transaction costs for services that will improve their financial and management capacities. Provision of basic industrial infrastructure and facilities such as energy will also be less cost-effective at those micro-scales. There will be the need to merge, consolidate or organize particularly the micro-scale firms into cooperatives to improve economies of scale. The district assemblies and municipalities with several micro-scale and very small-scale firms should establish industrial clusters with improved building structures, processing equipment (either for rent, lease of purchase) and improved access to utilities such as water and energy (decentralized and diversified with user fees or tariffs). As noted in Chapter 4, merger or consolidation of firms in Eastern region will require careful scrutiny and analysis of post-merger effect on concentration. However, most of the food crops agro-processing firms in the other regions can merge or be consolidated without unduly affecting prices and post-merger concentration.

13.3.4 Improve access to modern technologies and facilities.

As noted in the recommendation above, it will be more cost-effective to improve access to modern technologies if the micro-scale and the very small-scale firms are organized into industrial clusters where for example the transaction costs for lease finance and operational lease for improved processing facilities could be drastically reduced. It could also provide the opportunity to pull the risks of the several micro-scale firms and improve their financial capacity for acquiring improved processing, storage or packaging facilities. Modern processing, packaging and storage technologies developed in other emerging countries could be adopted and adapted for use in Ghana without 'reinventing the wheel'. MSMEs could be exposed to recently developed technologies through organizing workshops, seminars, exhibition and mini technological fairs.

13.3.5 Improve awareness and capacity for quality control and standardization.

There is a critical need for enforcement of food quality standards in the country. There is also the need to improve the awareness, knowledge and involvement of food crops agro-processing firms in policy and standard formulation and implementation. It was noted from the study that although the GSA has standards for food processing, most of the firms are not in compliance. There will be the need for the GSA and other relevant institutions to enhance awareness and compliance to these standards in order to improve the quality and competitiveness of processed agro products from Ghana. The enforcement mechanisms should have complementary education and capacity development activities to help the food crops agro-processing firms apply the knowledge to improve the quality of their products. There could also be award and other incentive systems with enhanced quality control and compliance with the standards as some of the major criteria for winning the award and recognition. The research findings also indicate that for food crops agro-processing firms to improve their competitiveness, they will have to enhance their capacity for packaging, promoting and marketing quality products.

13.3.6 Enhance entrepreneurial and managerial skills.

The research findings show that females constitute the greatest proportion of food crop agro-processing owners. Another characteristic of the owners of most food crop agro-processing firms is their low level of formal educational. As noted in the discussions, this will affect their training and skills acquisition as well as information search and adoption of technology. The research findings also indicate that level of education is one of the major factors determining access to credit. Moreover, over 80 percent of the firm owners interviewed lacked adequate capacity in areas such as resources mobilization, operational and financial management.

This finding should underlie the need for periodic training and other capacity development activities for agro-processing firms in order to improve their operational and managerial efficiency. Currently, most extension services provided by MOFA officers focus primarily on farming practices. As noted above, extension officers should be provided with the capacity to assist MSMEs in business planning, business operation and management, marketing and financial management.

Training manuals and audio visual materials should be developed and used to train extension officers and relevant NGOs on how to assist MSMEs in for example, financial records keeping, business planning, resources management (including human resources), operational efficiency and profitability planning, marketing, products and market segmentation, order sourcing, scheduling and sales management. The audio-visual training materials should be translated into Akan, Ewe and

Hausa for ease of comprehension when used for MSMEs who may not have adequate capacity to understand the issues in English.

The training activities should also use literacy and numeracy techniques to enhance the capacity of the MSMEs to understand and apply the knowledge and skills from the training activities. Since women are the dominant stake holders in this subsector, there will be the need for capacity enhancement activities that target female firm owners.

13.4 National Implementation of the Strategic Intervention Framework.

The implementation of the strategic intervention framework will require multi-stakeholder commitment, collaboration and coordination to mutually enhance the outcomes from the diverse interventions from relevant institutions. Figure 13.3 presents the implementation of the strategic intervention framework.

13.4.1 Organize national communication and visibility forums.

There should be extensive communication and visibility activities through-out the nation to improve awareness of the key plans and practices that should be taken collaboratively to transform food crops agro-processing in the nation into a billion-dollar industry. Activities should include panel discussions in the media that will focus on for example, the specific financial needs of SMEs in agro-processing of food crops, their credit characteristics, the opportunities, risks, constraints and relevant measures (including policy, regulation and incentives) for developing innovative financial products and services targeting their investment, financing and risk management needs.

Relevant findings from this book could be shared at some of these media panel discussions. This will be important because, several studies published in books and scientific journals sometimes do not reach the appropriate end-users. There is also the problem with the capacity of most policy and decision makers to understand and apply the knowledge due partly to the scientific and overly technical nature of the publication. This is why it is useful to apply simplified communication and dissemination means to help improve the utility of the science-based information.

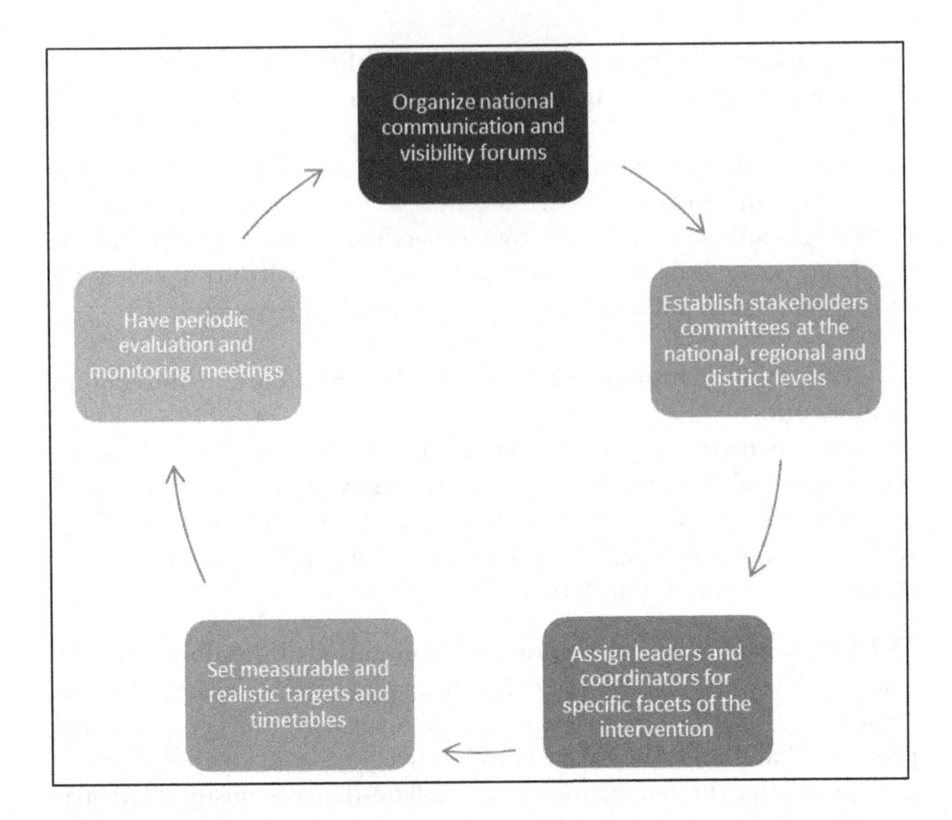

There should be stakeholder meetings in the different regions to discuss key issues in the respective regions and specific measures that should be taken to address those issues. UNU-INRA organized a stakeholders' seminar in December 4th, 2013 to share the major findings of this research and to provide the platform for dialogue. The seminar was attended by key stakeholders including the relevant ministries, EDAIF, VCTF, MASLOC, NBBSI and representatives from banks such as ECOBANK and ADB. There were also representatives from international development agencies and multilateral development banks such as SNV-Ghana and IFC. The proceedings of this seminar really enriched the discussions in this book. Such seminars should be encouraged throughout the nation.

There should be audio-visual materials in major languages. These materials should apply tools such as functional literacy and numeracy to enhance understanding of the presentations. The presentations in the audio-visual materials should feature for example, how the firms could improve their financial management, record keeping, liquidity, operational efficiency, profitability and general access to market-based financial products such as debt and equity financing, lease finance and insurance. The audiovisual materials should be distributed on flash drives to key stakeholders.

13.4.2 Establish stakeholders committees at the national, regional and district levels

The communication and visibility activities should be used to establish stakeholders committees at the national, regional and district levels. These stakeholder committees should have representation from government, financial institutions, SMEs involved in food-crops agro-processing, development agencies and other non-governmental organizations. At each level of the committee, appropriate policy and regulatory framework should be developed to provide guidance, oversight, relevant incentives and to address key bottlenecks to improving food crops agro-processing at that level of policy- and decision-making.

13.4.3. Assign leaders for each facets of the intervention at the various stakeholders' committee levels.

At each level of policy- and decision-making, leaders with relevant expertise should be assigned to oversee the strategic implementation of specific facets of the intervention. For example, representative from EDAIF could be assigned to oversee the implementation of measures to facilitate access to finance and investment products and services to food crops agro-processing firms. Representative from Ministry of Trade and Industry could oversee the implementation of consolidation and establishment of industrial clusters to pull micro-scale firms together and to achieve viable scale of operation. Representative from Ghana Standards Authority could oversee implementation of measures to improve products standards and quality control.

13.4.4 Set specific, measurable and realistic targets and timetables.

Each stakeholders committee should set up specific tools and indicators with targets and timetables for measuring progress in relevant policy development and implementation at their level of policy- and decision-making. For example, at the district level, stakeholder committees can set up specific outputs such as the number of financial institutions that should have reasonable credit terms, the number of MSMEs that should have access to these credits, the target amount in Ghana Cedis that should be disbursed to MSMEs, the number of micro or very small-scale firms that are organized into cooperatives or number of processing cooperatives established, and the number of industrial clusters with processing, storage, packaging and energy facilities in operation in the district at specific timeline. It will also include other measurable targets such as number of SMEs trained for example in financial record keeping, quality control and business management at specific timeline.

13.4.5 Have periodic evaluation and monitoring meetings.

The stakeholder committees should organize periodic meetings to evaluate performance and achievements of specific outputs and outcomes. The meeting should be used also to identify key risks and constraints and measures to address specific skills and to enhance performance.

13.5. Conclusion

This chapter has discussed a simple framework to prioritize issues and has also provided some actionable steps for implementation of the key interventions suggested in the book. Among the key issues identified for intervention are the need to develop innovative and inclusive financial products and services customized to meet the specific needs of MSMEs in food crops agro-processing. The financial products and services should be comprehensive and complementary involving a credit facilities, equity finance, lease and vendor finance, insurance, business and financial advisory services. There should be tax incentives such as tax rebates to firms that put in place measures to improve quality standards and achieve specified production and efficiency benchmarks.

The predominance of micro-scale firms makes it imperative to merge, consolidate or organize them into cooperatives to improve economies of scale. Industrial clusters with improved building structures, processing equipment, storage, transportation, packaging facilities and utilities such as water and energy should be provided to these cooperatives to improve their operations, products and financial viability. Such arrangements should be pursued to provide the opportunity to pull the risks of the several micro-scale firms to improve their financial capacity for acquiring improved processing, storage or packaging facilities.

Enforcement of standards and quality control should be improved and there should be extensive education and other capacity development programs to improve compliance to standards and other measures to improve product quality and assurance.

The chapter has also discussed national strategic implementation plan for pursuing the key interventions suggested. It is only through implementation that Ghana can realize the potential of transforming food crops agro-processing firms into competitive and viable firms to pursue national efforts at import substitution such as the measures for controlling the importation of rice. It is important to stress that the implementation of the strategic intervention framework will require multi-stakeholder commitment, collaboration and coordination to mutually enhance the intended outputs and outcomes. Stakeholders should pursue nation-wide communication and visibility activities to improve awareness and commitment to measures in order to promote food crops agro-processing in the nation. Stakeholder meetings should be used to establish committees at different levels of policy- and decision-making to pursue the strategic implementation of the measures across the nation.

Specific tools and indicators with targets and timetables should be developed by stakeholder committees to measure, monitor and report achievements of expected outputs and outcomes.

It is hoped that such practical measures should move the nation forward in its efforts at promoting agro-processing and agribusinesses in order to sustain initiatives aimed at controlling importation of cheap products and also to prevent the decline of industrialization in the nation.

REFERENCES

GSS. (2011). *Contributions to GDP by Sectors. Cited in MOFA Facts and Figures 2010.* Accra: Ghana Statistical Services.

(GIPC), G. I. (n.d.). *Food production and processing.* Retrieved March 21st, 2013, from http://tudec.org/downloads/Ghana_Investment_Opportunities/Food.pdf.

Abor, J., & Quartey, P. (2010). Issues in SMEs Development in Ghana and South Africa. *International Journal of Financial Economics, 39*, 218-228.

Abott, J. (1994). Agricultural processing enterprises development potentials and links to the small holder. In J. Von Braun, & J. Eileen Kennedy, *Agricultural Commercialization, Economic Developement and Nutrition* (pp. 153-165). Baltimore: Johnss Hopkins University Press.

Akpan, S. B., Inimfon, P. V., Udoka, S. J., Offiong, E. A., & Okon, U. E. (2013). Determinants of Credit Access and Demand among Poultry Farmers in Akwa Ibom State, Nigeria. *American Journal of Experimental Agriculture, 3*(2), 293-307.

Akudugu, M. (2012). Estimation of the Determinants of Credit Demand by Farmers and Supply by Rural Banks in Ghana's Uppe East Region. . *Asian Journal of Agriculture and Rural Development, 2*(2), 189-200.

Albersmeier, F., Schulze, H., & Spiller, A. (2010). System Dynamics in Food Quality Certification: Development of an Audit Integrity System. *International Journal of Food System Dynamics, 1*, 69-81.

Aryeetey, E. (1996). *The Formal Financial Sector in Ghana after the Reforms.* Overseas Development Institute.

Aryeetey, E., & Ahene, A. (2005). *Changing Regulatory Environment for Medium Size Enterprises and Their Performance in Ghana.* Centre on Regulation and Competition.

Asuming-Brempong, S. (2003). Economic and Agricultural Policy Reforms and their Effects on the Role of Agriculture in Ghana. *Roles of Agricltrure Project International Conference.* Italy, Rome: FAO.

Ata, J. (1974). *Processing and quality Characteristics of GHanaian Commercial Palm Oils.* Accra: Food Research Institute.

Aworh, C. O. (2008). The Role of Traditional Food Processing Technologies in National Development: the West African

Experience. In G. Robertson, & Lupien, J.R., *Using Food Science and Technology to Improve Nutrition and Promote National Developmnet.* International Union of Good Science and Technology.

Barbieri, C., & Mshenga, P. M. (2008). The Role of the Firm and Owner Characteristics on the Performance of Agritourism Farms. *Sociologia Ruralis, 48*(2), 166-183.

Boundless. (2012). *Business / Financial Statements / Ratio Analysis and Statement Evaluation* . Retrieved August 13th, 2013, from Boundless.com: https://www.boundless.com/business/financial-statements/ratio-analysis-and-statement-evaluation/comparisons-within-an-industry/

Brunso, K., Fjord, T., & Grunert, K. (2002). *Consumers' Food Choice and Quality Perception.* Aarhus School of Business, Denmark.

Business News. (2013, December 2). *Ghana's public debt hits GHC46.1 billion.* Retrieved December 2, 2013, from Business news: http://www.ghanaweb.com/GhanaHomePage/business/artikel.php?ID=293888

Business News. (2013, December 12). *Inflation shoots up to 13.2%.* Retrieved December 12, 2013, from Business News.com: http://www.ghanaweb.com/GhanaHomePage/business/artikel.php?ID=294916

Buzby, J. (2003). *International Trade and Food Safety: Economic Theory and Case Studies, Agricultural Economics Report.* Washington D.C: U.S. Department of Agriculture, .

Cacciolattie, L., & Wan, T. (2012). A Study of Small Business Owners' Personal Characteristics and the Use of Marketing Information in the Food and Drink Industry: A Resource-Based Perspective. *International Journal on Food System Dynamics, 3*(2), 171-184.

Carmines, E. G., & Zeller, R. A. (1979). *Reliability and validity assessment. Quantitative applications in the social sciences.* Beverly Hills. CA: Sage Publications.

Chandrasekhar, K., Sreevani, S., Seshapani, P., & Pramodhakumari, J. (2012). A Review on Palm Wine. *International Journal of Research in Biological Sciences*(2249-9687), 33-38.

Cramer, C. (1999). Can Africa Industrialize by Processing Primary Commodities? The Case of Mozambican Cashew Nuts. *World Development, 27*(7), 1247-1266.

Cui, Q. (2013). Cause Analysis of Non-agricultural Trend of Credit Cooperatives and Reversion Approaches. *Asian Agricultural Research, 5*(4), 22-25.

Daily Graphic. (2013, March 21). *AGRICULTURE:World Bank, USAID Support Agriculture.* Retrieved August 28, 2013, from Ghana Distircts.com: http://www.ghanadistricts.com/news/?read=48725

Daily Graphic. (2013, November 6). *Businesses cry out to trade minister.* Retrieved November 11, 2013, from Daily Graphic: http://www.ghanaweb.com/GhanaHomePage/business/artikel.php?ID=291201

Dalitso, K., & Quartey , P. (2000). *The Policy Environment for Promoting Small and Medium-Sized Enterprises in Ghana and Malawi.* . Finance and Research Working Paper Series Paper No 15.

Davis, B., Winters, P., Carletto, G., Covarrubias, K., Quinones, E. J., Alberto, Z., . . . Digiuseppe, S. (2010). A Cross-Country Comparison of Rural Income Generating Activities. *World Development, 38*(1), 48-63.

Derbile, E. K., Abubakari, A., & Dinye, R. D. (2012, November 28). Diagnosing Challenges of Small-Scale Industries Ghana: A Case of Agro-Processing Industries in KassenaNankana District. *African Journal of Business Management, 6*(47), 11648-11657.

Derbile, E. K., Abubakari, A., & Dinye, R. D. (2012). Diagnosing challenges of small-scale industries in Ghana: A case of agro-processing industries in Kassena-Nankana District. *African Jounal of Business Management*, 11648-11657.

DFID. (1998). *Improving the competitiveness and Marketability of Locally-Produced Rice in Ghana.* Department for International Development.

Dijkstra, T. (2001). *Agricultural exports and economic growth in less developed courntries.* Netherlands: Working paper the African studies centre.

Drake, P. P. (2001). *Financial ratio Analysis.* Retrieved May 24th, 2013, from http://educ.jmu.edu/~drakepp/principles/module2/fin_rat.pdf

E., S. (1998). *Post-harvest loss due to pests in dried cassava chips and comparative methods for its assessment. A case study on small-scale farm households in Ghana.* Eschborn: GTZ.

Earle, R. (1983). *Unit Operations in Food Processing.* Toronto: Pergamon Press.

EDIF. (2009). *Export Development and Investment Fund. Annual Report*.

EDIF. (2011). *Agriculture and Agro-Processing Development and Credit Facility.* Retrieved June 27th, 2013, from http://www.edifgh.org/en/facilities-of-the-fund/agriculture-and-agro-processing-development-credit-facility.php

Edwards, C. (1968, October). The Meaning of Quality. *Quality Progress*, pp. 36-39.

Eze, C. C. (2003). Resource Use Efficency in Cocoyam Production in Ihitte/Uboma Area of Imo State, Nigeria. *Journal of Applied Sciences, 6*(3), 3865-3871.

FAO. (1985). *FAO Training Series No. 17/1. Volume II .Training manual on the prevention of post-harvest food losses.* Rome: UN FAO.

FAO. (1997). *The State of Food and Agriculture.* Retrieved April 4th, 2013, from http://www.fao.org/dorcrep/w5800e/5800e00.htm

FAO and UNIDO. (2010). *African Agribusiness and Agro-industries Development Initiative: A programme framework (3ADI).* Rome: UN FAO and UNIDO.

Frewer, L., Scholderer, J., & Lambert, N. (2003). Consumer Acceptance of Functional Foods. *British Food Journal, 105*(5), 714-731.

GAIP. (2012). *GAIP Newsletter.* Retrieved June 27th, 2013, from http://www.gaip-info.com/wp/wp-content/uploads/GAIP_NEWS_Dec2012.pdf

Garvin, D. (1988). *Managing Quality. The Strategic and Competitive Edge.* New York: The Free Press.

Gawron, J., & Theuvsen, L. (2009). *Agrifood Certification Schemes in an Intercultural Context: Theoretical Reasoning and Empirical Findings.* European Association of Agricultural Economists.

Gbedemah, F. (n.d.). Certification in Manufaction Companies in Ghana: Prospects and Challenges. *A Master of Science (MSc) thesis submitted to the Lund University.*

Ginder, R. G., & Artz, G. M. (2001). *Financial Standards for Iowa Agribusiness Firms.*

GNA. (2013, August 05). *AGOTIME Z:Ziope 'cries' for tomato factory.* Retrieved August 10, 2013, from Ghana Districts.com: http://www.ghanadistricts.com/news/?read=49878

GNA. (2013, September 11). *Go for EDAIF loan - MOTI urges entrepreneurs.* Retrieved November 8, 2013, from Ghana News Agency:

http://www.ghanaweb.com/GhanaHomePage/business/artikel.php?ID=285491

GNA. (2014, January 3). *Ghana eases ban on rice importation.* Retrieved January 10, 2014, from Ghana Business News.com: http://www.ghanabusinessnews.com/2014/01/03/ghana-eases-ban-on-rice-importation/

GoG. (2010). *Micro Finance and Small Loans Center (MASLOC).* Retrieved June 27th, 2013, from http://www.presidency.gov.gh/our-government/agencies-commissions/micro-finance-and-small-loans-center

Goni, M. M., & Baba, B. A. (2007). Analysis of Resource-Use Efficiency in Rice Production in the Lake Chad Area of Borno State, Nigeria. *Journal of Sustainable Development, 3*(2), 31-37.

Government, U. (2011, February). *Feed the Future.* Retrieved March 24th, 2013, from http://feedthefuture.gov/sites/default/files/country/strategies/files/GhanaFeedtheFutureMulti-YearStrategy_2011-08-03.pdf

Greasley, A. (2007). *Operations Management.* Los Angeles: Sage Publications.

Grunert, K. (2002). Current Issues in the Understanding of Consumer Food Choice. *Trends in Food Science and Techonology, 13*, 275-285.

GSS. (2008). *Ghana Living Standards Survey.* Accra: Ghana Statistical Services.

GSS. (2011). *Contributions to Gross Domestic Product by Sector in Ghana. Accra.* Accra: Ghana Statistical Services.

Gujarati, N. D. (2004). *Basic Econometrics.* (4th Edition ed.). UK: Mcgraw Hill.

Gumert, K. (2003). How Changes in Consumer Behavior and Retailing Affect Competence Requirements for Food Producers and Processors. *Global Markets for High-Value Foods.* U.S. Department of Agriculture, Economic Research Service.

Hagerbaume, J. B. (1977, August). The Gini Concentration Ratio and the Minor Concentration Ratio: A Two Parameter Index of Inequality. *The Review of Economics and Statistics, 59*(3), 357-379.

Hatanaka, M., Bain, C., & Busch, L. (2005). Third-Party Certification in the Global Agrifood System. *Food Policy, 30*, 354-369.

Heathfield, D. F. (1971). *Production Functions, Macmillan studies in economics.* New York: Macmillan Press.

Hooker, N., & Caswell, J. (1999). A Framework for Evaluation Non-Tariff Barriers to Trade Related to Sanitary and Phytosanitary Regulation. *Journal of Agricultural Economics, 50*(2), 234-246.

Hossain, M. Z., & Al-Amri, K. S. (2010). Use of Cobb-Douglas production model on some selected manufacturing industries in Oman. *Education, Business and Society: Contemporary Middle Eastern, 3*(2), 78-85.

Ibarz, A., Barbosa-Canovas, & Gavarin, V. (2003). *Unit Operations in Food Engineering.* Boca Raton, Fla.

IFAD. (2013, August 18). *Rural and Agricultural Finance Programme (RAFiP).* Retrieved August 18, 2013, from International Fund for Agricultural Development: http://operations.ifad.org/web/ifad/operations/country/project/tags/ghana/1428/project_overview

Islam, A. M., Khan, A. M., Obaidullah, M. A., & Alam, S. M. (2011, March). Effect of Entrepreneur and Firm Characteristics on the Business Success of Small and Medium Enterprises (SMEs) in Bangladesh. *International Journal of Business and Management, 6*(3), 289-299.

Jaffee, S. M., & Spencer, H. (2001). Agro-food Exports from Developing Countries: The Challenges Posed by Standards in Global Agricultural Trade and Developing Countries. In M. Ataman, Aksoy, & Beghin, *Global Agricultural Trade and Developing Courntries* (pp. 91-114). Washington DC: The World Bank.

John, G., Schramm, M., & Spiller, A. (2005). *Institutional Change in Quality Assurance: The Case of Organic Farming in Germany.*

Karipidis, P., Athanassiadis, K., Aggelopoulos, S., & Giompliakis. (2009). *Factors Affecting the Adoption of Quality Assurance Systems in Small Food Enterprises.*

Kipene , V., Lazaro , E., & Isi, A. C. (2013). Labour Productivity Performance of Small Agro-Processing Firms in Mbeya and Morogoro, Tanzania. *Journal of Economics and Sustainable Development, 4*(3), 125-134.

Koenig, S. R., & Doye, D. G. (1999). *Agricultural Credit Policy.* USDA.

Kotler, P. (2007). Global Standardization - Courting Danger. *Emerald Backfiles, 3*(2).

London Economics. (2007). *Structure and performance of six European wholesale electricity markets in 2003, 2004 and 2005.* London: London Economics. Retrieved from http://ec.europa.eu/competition/sectors/energy/inquiry/electricity_final_part1.pdf

Mamood, A. N., Khalid, M., & Kouser, S. (2009). The Role of Agricultural Credit in the Growth of Livestock Sector: A Case Study of Faisalabad. *Pakistan Vetinary Journal, 29*(2), 81-84.

MASLOC. (2010). *Management of MASLOC Targets Agriculture Sector for Loans Disbursement.* Retrieved June 27th, 2013, from http://www.masloc.gov.gh/2/12/MASLOC-News.html?item=26

McNeill, M. (2005). Ergonomics in Post-harvest Agro Processing. *African Newslett on Occupational Health and Safety, 15*(1), 11-13.

MLGRD. (2013, June 15). Investment and Business Potential of Eastern Region, Ministry of Local Government and Rural Development. Accra, Greater Accra, Ghana. Retrieved from http://www.ghanadistricts.com/districts/?r=4&_=66&sa=496

MoFA. (1997). *Agriculture in Ghana, Facts and Figures.* Accra, Ghana: Ministry of Food and Agricultre.

MoFA. (2002). *Food and Agriculture Sector Development (FASDEP).* Accra: Government of Ghana.

MoFA. (2002). *Food and Agriculture Sector Policy (2002).* Government of Ghana.

MoFA. (2007). *Food and Agriculture Sector Development Policy (FASDEP II).* Accra: Ministry of Food and Agriculture.

MoFA. (2010). *Agriculture in Ghana, Facts and Figures.* Accra: Ministry of Food and Agriculture.

MoFA. (2010). *Agriculture in Ghana, Facts and Figures.* Accra, Ghana: Ministry of Food and Agriculture.

MoFA. (2011). *Agriculture in Ghana, Facts and Figures 2010.* Accra: Government of Ghana.

MOFA. (2012). *Crop Production Estimates 2011 and 2012.* Accra: Ministry of Food and Agriculture. Retrieved April 4, 2012, from http://mofa.gov.gh/site/?page_id=5889

Mugera, A. W. (2012). Sustained Competitive Advantage in Agribusines: Applying the Resource-Based Theory to Human Resources. *International Food and Agribusiness Management Review, 14*(4), 27-48.

Myjoyonline. (2013, July 03). *Fruits Ban: Agric Ministry admits Ghana can't meet all US standards.* Retrieved August 28, 2013, from Ghana Districts.com: http://www.ghanadistricts.com/news/?read=49444

National Development Planning Commission, N. (2005). *Growth and Poverty Reduction Strategy.* Accra, Ghana: NDPC.

NDPC. (2005). *Growth and Poverty reduction Strategy (GPRS II).* Accra: Government of Ghana.

Nimoh, F., Tham-Agyekum, E. K., & Nyarko, P. K. (2012). Resource Use Efficiency in Rice Production: the Case of Kpong Irrigation Project in the Dangme West District of Ghana. *International Journal of Agriculture and Forestry, 2*(1), 35-40.

Nto, P. O., & Mbanasor, J. A. (2011). Productivity in agribusiness firms and its determinants in Abia State, Nigeria. *Journal of Economics and International Finance, 3*(12), 662-668.

ODI. (2005). The Future of Small Farms. *Proceedings of a Research Workshop.* London: Imperial COllege.

Ofei, K. A. (2004). Terms and Access to Credit: Perceptions of SME/Entrepreneurs in Ghana. *International Conference on Ghana at Half Century.* Accra: ISSER/Cornell University.

Okerenta, S. I., & Orebiyi, J. S. (2005). Determinants of Agricultural Credit Supply to Farmes in the Niger Delta Area of Nigeria. *Journal of Agriculture and Social Research, 5*(1), 67-72.

Okorley, E. L., & Ayekpa, E. (2010). Factors that can Influence contract farming partnership betwen Farmers and Agro-Processing Industries. A Ghanaian Case Study. *The Journal of Business and Enterprise Development,* 41-50.

Okorley, E. L., Zinnah, M. M., Mensah, A. O., & Owens, M. (1999). *Women in Agro-processing in Ghana: A Case Study of the State of Women in Small-Scale Fish Smoking in the Central Region of Ghana.*

P., J. (198). *Food Technology.* Ullmann's Encyclopedia of Industry.

Plucknett, D. L. (1979). *Small-scale processing and storage of tropical root crops.* Boulder, Colorado: Westview.

Ponte, S., & Gibbon, P. (2005). Quality Standards, Conventions and the Governance of Global Value Chains. *Economy and Society, 34*(1), 1-31.

Robinson , E., Shashidhara, K., & Xinshen, D. (2012). *Food Processing and Agricultural Productivity Challeges: The Case of Tomatoes in Ghana.* IFPRI.

Robinson, E., Kolavalli, S., & Diao, X. (2012). *Food processing and Agricultural Productivity challenges: The case of Tomatoes in Ghana.* International Food Policy Research Institute.

Rolle, R. (2006). Processing of Fruits and Vegetables for Reducing Postharvest Losses and Adding Value. In R. Rolle, *Postharvest Management of Fruit and Vegetables in the Asia-Pacific Region* (pp. 32-42). Tokyo: Asian Productivity Organization (APO).

Stanley, J. (1971). Reliability. In R. L. Thorndike, *Educational Measurement, (2nd ed)* (pp. 356-442). Washington DC: American Council on Education.

Storey, D. (1994). *Understanding the Small Business Sector.* London: Routledge.

Theuvsen, L., & Plumeyer, C. (2008). *Certification Schemes, Quality-Related Communication in Food Supply Chains - Consequences for IT-Infrastructures.*

UNIDO, U. N. (2008). *The Importance of Agro-Industry for Socio-econmic Development and Poverty Reduction.* New York.

V., S. (2006). *Supply Chain, Power Relationships and Local Food Systems.* Federico: Department of Agricultural Economics, University of Naples.

VCTF. (2013). Venture Capital Trust Fund - Operations. Accra: Venture Capital Trust Fund. Retrieved January 28, 2013, from http://www.venturecapitalghana.com.gh/AboutUs/Operations/tabid/95/Default.aspx

Vermeulen, H., & Bienabe, E. (2007). *What About The Food, Quality Turn in South Africa? Focus on The Organic Movement Development.*

Vos, E. (1997). *Financial Analysis and Interpretation.* Retrieved May 23rd, 2013, from http://wms-soros.mngt.waikato.ac.nz/NR/Personal/Ed%20Vos/includes/publications/Textbookpdf/chap7p.pdf

White, S. (1999). *Women's Employment in the Agro and Food Processing Sector: South Asia and East Africa.* Canada: Aga Khan Foundation.

Wirth, F., Stanton, L. J., & Wiley, B. J. (n.d.). The Relative Importance of Search Versus Credence Product Attributes: Organic and Locally Grown. *Agricultural and Resource Economics Review, 40*(1), 48-62.

Wongnaa, C. A., & Ofori, D. (2012). Resource-use Efficiency in Cashew Production in Wenchi Municipality, Ghana. *Agris on-line Papers in Economics and Informatics, 4*(2).

World Bank. (2010). *Missing Food: The Case of Postharvest Grain Losses in Sub-Saharan Africa.* Washington DC: World Bank Group.

Appendix 1

Location	Statistics	Q	LAB	CAP	RAW	ENE	TEC
eastern	mean	840.835	7791.667	2561.242	240.6825	1102.911	0.625
	max	843.88	30000	2562.98	242.65	2500	1
	min	840	500	2560	240.11	45	0.5
	skewness	1.862624	0.966374	0.45765	2.46215	0.554921	1.154701
	kurtosis	5.33218	2.822161	1.826382	8.218987	4.410546	2.333333
	sd	1.143277	9932.633	0.994905	0.65401	586.9275	0.226134
	N	12	12	12	12	12	12
Central	mean	223834.9	3557.175	3247.3	6519.642	1212.634	0.539683
	max	1.23E+07	31440	68360	53760	1920	1
	min	840.11	192	72	240.11	1201.224	0.5
	skewness	7.743284	3.00204	6.530968	3.267915	7.747008	3.112267
	kurtosis	60.97825	14.39171	47.74453	14.99	61.01613	10.68621
	sd	1550152	5286.479	8871.131	9715.536	90.5572	0.13624
	N	63	63	63	63	63	63
northern	mean	19409.76	1791.654	753.8243	13224.34	1210.811	0.560976
	max	115200	30000	2560.795	113750	13440	1
	min	840.11	192	96	240.11	72	0.5
	skewness	2.718902	5.930852	1.573485	2.960856	5.694343	2.310604
	kurtosis	9.911451	37.11901	4.792928	12.9987	35.21567	6.338889
	sd	24048.79	4572.519	613.8467	21606.28	2008.725	0.165647
	N	41	41	41	41	41	41
Brong Ahafo	mean	26224.2	3728.9	1584.675	8935.62	1354.399	0.55
	max	192000	36960	9600	57600	8840	1
	min	840.11	360	35	240	120	0.5
	skewness	2.888742	3.192595	2.111562	2.017734	3.646754	2.666667
	kurtosis	13.66401	14.53258	7.330212	8.504725	19.45441	8.111111

Source: Field Survey 2012

Appendix 2

Correlation Matrix for Eastern Region variables

	LnQ	LnLAB	LnCAP	LnRAW	LnENE	LnTEC
LnQ	1					
LnLAB	0.2757	1				
LnCAP	-0.2902	0.0885	1			
LnRAW	-0.1095	-0.3314	-0.2784	1		
LnENE	0.2491	-0.2292	0.1458	0.2467	1	
LnTEC	0.4438	0.6939	-0.0152	-0.2268	0.3533	1

Correlation Matrix for Central Region variables

	LnQ	LnLAB	LnCAP	LnRAW	LnENE	LnTEC
LnQ	1					
LnLAB	0.3349	1				
LnCAP	0.2177	0.3984	1			
LnRAW	0.5289	0.3888	-0.0566	1		
LnENE	0.0999	-0.0937	0.0099	0.0497	1	
LnTEC	0.1431	0.5028	0.1518	0.1083	-0.0373	1

Correlation Matrix for Northern Region variables

	LnQ	LnLAB	LnCAP	LnRAW	LnENE	LnTEC
LnQ	1					
LnLAB	-0.3835	1				
LnCAP	0.0128	0.2457	1			
LnRAW	0.4954	0.2214	0.221	1		
LnENE	-0.0654	0.359	0.3679	0.2139	1	
LnTEC	0.044	0.3273	0.3825	0.3666	0.4661	1

Correlation Matrix for Brong Ahafo Region variables

	LnQ	LnLAB	LnCAP	LnRAW	LnENE	LnTEC
LnQ	1					
LnLAB	0.6706	1				
LnCAP	0.4374	0.4275	1			
LnRAW	0.7196	0.5758	0.402	1		
LnENE	0.2253	0.4167	0.2978	0.3363	1	
LnTEC	0.3493	0.3853	0.1882	0.3866	0.1588	1

Correlation Matrix for Western Region variables

	LnQ	LnLAB	LnCAP	LnRAW	LnENE	LnTEC
LnQ	1					
LnLAB	0.501	1				
LnCAP	0.4131	0.2499	1			
LnRAW	0.6013	0.4756	0.4685	1		
LnENE	0.1445	0.3026	0.1207	0.1327	1	
LnTEC	0.2745	0.432	0.005	0.2538	-0.1331	1

Source: Field Survey 2012

Printed in the United States
By Bookmasters